P9-DET-350

Darwin

'Tim Lewens' new book is a masterclass in the interpretation of Darwin. Anyone wanting to know what Darwin meant and how his words bear on the great debates of our time will find here a rare combination of clarity, learning, and fresh thinking. Indispensable.'

Gregory Radick, University of Leeds

'A clear, well-written, fair, broad-ranging and student-friendly introduction to Darwinian thinking'

Kim Sterelny, Victoria University of Wellington and Australian National University

'Charles Darwin has been enlisted to support many different, and often contradictory, intellectual and political agendas. This is a clear and careful philosophical examination of which of these causes Darwin is willing and able to support. An excellent introduction to Darwin's intellectual orientation and the implications of his thought.'

Paul Griffiths, University of Queensland, Australia

'Charles Darwin remains as influential as ever. He is a hate figure of the religious right which only adds to his lustre in the eyes of everybody else. Tim Lewens brilliantly explores the extraordinary role that Darwin has played not only in science and philosophy but also right across the full range of human affairs. Lewens' book contradicts the belief that nothing more that is fresh and interesting could be added to all the existing writings about Darwin.'

Sir Patrick Bateson, University of Cambridge, UK

Routledge Philosophers

Edited by Brian Leiter
University of Texas, Austin

Routledge Philosophers is a major series of introductions to the great Western philosophers. Each book places a major philosopher or thinker in historical context, explains and assesses their key arguments, and considers their legacy. Additional features include a chronology of major dates and events, chapter summaries, annotated suggestions for further reading and a glossary of technical terms.

An ideal starting point for those new to philosophy, they are also essential reading for those interested in the subject at any level.

Hobbes	A. P. Martinich
Leibniz	Nicholas Jolley
Locke	E. J. Lowe
Hegel	Frederick Beiser
Rousseau	Nicholas Dent
Schopenhauer	Julian Young
Freud	Jonathan Lear
Kant	Paul Guyer
Husserl	David Woodruff Smith
Darwin	Tim Lewens

Forthcoming:

Aristotle	Christopher Shields
Spinoza	Michael Della Rocca
Hume	Don Garrett
Fichte and Schelling	Sebastian Gardner
Rawls	Samuel Freeman
Merleau-Ponty	Taylor Carman
Heidegger	John Richardson

Tim Lewens

Darwin

Routledge
Taylor & Francis Group

LONDON AND NEW YORK

First published 2007
by Routledge
2 Park Square, Milton Park, Abingdon, Oxon OX14 4RN

Simultaneously published in the USA and Canada
by Routledge
270 Madison Ave, New York, NY 10016

Routledge is an imprint of the Taylor & Francis Group, an informa business

© 2007 Tim Lewens

Typeset in Joanna MT and Gills Sans by Taylor & Francis Books
Printed and bound in Great Britain by MPG Books Ltd, Bodmin

All rights reserved. No part of this book may be reprinted or reproduced
or utilised in any form or by any electronic, mechanical, or other means,
now known or hereafter invented, including photocopying and recording, or
in any information storage or retrieval system, without permission in writing
from the publishers.

British Library Cataloguing in Publication Data
A catalogue record for this book is available from the British Library

Library of Congress Cataloging in Publication Data
Lewens, Tim.
Darwin / Tim Lewens.
p. cm. -- (Routledge philosophers)
Includes bibliographical references.
ISBN-13: 978-0-415-34637-5 (hardback : alk. paper)
ISBN-10: 0-415-34637-1 (hardback : alk. paper)
ISBN-13: 978-0-415-34638-2 (pbk. : alk. paper)
ISBN-10: 0-415-34638-X (pbk. : alk. paper) 1. Darwin, Charles, 1809-1882.
I. Title. II. Series.
B1623.L49 2006
192--dc22
2006014463

ISBN 10: 0-415-34637-1 (hbk)
ISBN 10: 0-415-34638-X (pbk)
ISBN 10: 0-203-59713-3 (ebk)

ISBN 13: 978-0-415-34637-5 (hbk)
ISBN 13: 978-0-415-34638-2 (pbk)
ISBN 13: 978-0-203-59713-2 (ebk)

For my mum and my dad

I suspect the endless round of doubts & scepticisms might be solved by considering the origin of reason. as gradually developed. see Hume on Sceptical Philosophy.

Charles Darwin, Notebook N

Chapter 59
Wave of Inventions

The reign of Queen Victoria was famous for the numerous discoveries and inventions which happened in it. One of the first of these was the brilliant theory of Mr Darwin propounded in his memorable works, *Tails of a Grandfather*, *The Manx Man*, *Our Mutual Friends*, etc. This was known as *Elocution* or the *Origin of Speeches* and was fiercely denounced in every pulpit.

W. C. Sellar and R. J. Yeatman, 1066 And All That

Acknowledgements

This book aims to cover a lot of ground, and as a result of that I needed a lot of help in writing it. Many of my friends and colleagues were generous enough to wade through the entire manuscript, and on several occasions they saved me from embarrassing errors and oversights. I am especially grateful to Patrick Bateson, Tony Bruce, David Buller, Emma Gilby, Paul Griffiths, Nick Jardine, Martin Kusch, Peter Lipton, Matteo Mameli, Greg Radick, Jim Secord, Kim Sterelny and three anonymous readers from Routledge. Every one of them took time to read the whole text closely, and to offer constructive comments on it. For discussion and comment on individual chapters I am indebted to André Ariew, Philippe Huneman, Mohan Matthen, Hugh Mellor, Staffan Müller-Wille and Denis Walsh. Tamara Hug, Steve Kruse, Dawn Moutrey and David Thompson provided invaluable logistical and moral support. Christina McLeish helped to produce a fine index. I also owe further thanks to Tony Bruce at Routledge for asking me to write this book, and to Kim Sterelny (whom Tony asked first), for turning the invitation down. Both Cambridge University and Clare College were kind enough to give me a term of sabbatical leave in the autumn of 2005, which enabled me to finish the project. Finally I am grateful to Emma Gilby, who not only made the manuscript better, but who made many other things better, too.

1809 Charles Robert Darwin born in Shrewsbury, on 12th February.

1817 Darwin's mother Susannah dies when Charles is aged eight.

1818 Attends Shrewsbury School. Later claims to have learned almost nothing from it.

1825 Begins studying medicine at the University of Edinburgh. He becomes friendly with Robert Grant, a follower of Lamarck.

1828 Bored by his lectures and sickened by surgery, Darwin abandons his medical training and arrives at Christ's College, Cambridge. His new chosen career is that of an Anglican priest.

1831 Graduates from Cambridge (no honours) and goes on his first geological field trip with Professor Adam Sedgwick. Later in the year his botanist friend Henslow arranges for him to travel as ship's naturalist on board a surveying ship, HMS *Beagle*. He sets sail on 27th December.

1832– The *Beagle* carries out its surveying business off the South
1835 American coast. Darwin makes numerous long trips to the interior. Visits the Galapagos Islands in September and October 1835.

1836 The *Beagle* returns to England, arriving at Falmouth on 2nd October.

1837 Moves to lodgings in Great Marlborough Street, London. Begins to make notes on 'transmutation'.

1838 Reads Malthus's *Essay on the Principle of Population*, which he later diagnoses as a key moment in the formulation of the principle of natural selection. Darwin begins to suffer from the illness that will plague him throughout the remainder of his life.

1839 Becomes a Fellow of the Royal Society. Marries his cousin, Emma Wedgwood. They move to Upper Gower Street. Darwin's *Journal of Researches* is published, now more usually known as *The Voyage of the Beagle*. The first of their ten children, William Erasmus Darwin, is born.

1842 The Darwins leave London and move to Down House in Kent. Darwin's *The Structure and Distribution of Coral Reefs* is published.

1844 Composes a substantial essay outlining his evolutionary theory, and gives instructions to Emma to arrange for its publication should his illness take his life. *Vestiges of the Natural History of Creation* is published (an evolutionary work written anonymously by Robert Chambers). It is a huge popular success, and is roundly criticised by Darwin's scientific peers.

1846 Begins an eight-year-long study of barnacles.

1848 Robert Darwin, Charles's father, dies.

1856 Starts to compose a book to be called *Natural Selection*. It is never completed.

1858 Receives an essay from Alfred Russel Wallace, which Darwin judges to contain 'exactly the same theory' as his own. A presentation of work by both Darwin and Wallace is staged at the Linnaean Society on 1st July. It is the first time Darwin's theory is aired publicly.

1859 Spurred by Wallace's essay, Darwin abandons *Natural Selection*, and instead produces a shorter 'abstract' of his theory, published in November as *On the Origin of Species*.

1862 Publishes his 'flank move on the enemy'. The enemy is natural theology, the work is *On the Various Contrivances by which British and Foreign Orchids are Fertilised by Insects*.

1868 Publishes *The Variation of Animals and Plants under Domestication*, a book which includes, among other things, Darwin's theory of inheritance.

1871 *The Descent of Man* is published – Darwin's primary effort to relate his theory to the human species.

1872 Publishes *The Expression of the Emotions in Man and Animals*, and the sixth and final edition of *The Origin of Species*.

1881 Darwin's last book appears, *The Formation of Vegetable Mould, Through the Action of Worms*.

1882 Dies on 19th April, and is buried in Westminster Abbey.

A Note on Texts

In an effort to encourage readers to consult Darwin's own works, the page references in this book are, wherever possible, to widely available editions of his major publications. I refer to four works with particular frequency, and for these works I use predictable abbreviations:

Origin refers to *The Origin of Species*, and unless otherwise stated all page references are to the 1985 Penguin Classics printing of the 1859 first edition (published with an introduction by J. W. Burrow).

Descent refers to *The Descent of Man*, and unless otherwise stated all page references are to the 2004 Penguin Classics printing of the 1877 second edition (published with an introduction by Adrian Desmond and James Moore).

Expression refers to *The Expression of the Emotions in Man and Animals*, and all page references are to the 1998 HarperCollins printing of the 1889 third edition (published with an introduction by Paul Ekman).

Autobiography refers to Darwin's initially unpublished recollections, and all page references are to the 2002 Penguin Classics edition entitled *Autobiographies* (with an introduction by Michael Neve and Sharon Messenger).

Introduction

'A Philosophical Naturalist'

I. DIAL 'M' FOR 'METAPHYSICS'

This book is a philosophical introduction to Darwin. It explores and evaluates the relevance of Darwin's thinking for The Big Questions – traditional philosophical questions about the mind, ethics, knowledge, politics and science. How can there be such a book? Darwin was not a philosopher, he was a natural historian. His published works are about coral reefs, climbing plants, barnacles, earthworms and orchids – they are not works of philosophy. Indeed, Darwin sometimes portrays himself as a philosophical airhead: 'My power to follow a long and purely abstract train of thought is very limited; I should, moreover, never have succeeded with metaphysics or mathematics' (*Autobiography*: 85).

 As a teenager, while staying at his uncle Josiah's house, Darwin met Sir James Mackintosh. Mackintosh's works are rarely read today, but he was a prominent philosopher of the time. Darwin was impressed by Mackintosh, and Mackintosh was impressed by Darwin. Darwin later tried to explain Mackintosh's good opinion: 'This must have been chiefly due to his perceiving that I listened with much interest to everything which he said, for I was as ignorant as a pig about his subjects of history, politics and moral philosophy' (*Autobiography*: 27).

Darwin eventually lifted his mind from the sty by reading several works of philosophy, albeit selectively. He came to Kant late in life, and the experience seems to have left him cold. But in the years immediately after the *Beagle* voyage, when he was still a young man, Darwin read David Hume and Adam Smith; he made extensive notes on works by Mackintosh, and other philosophers better known in Victorian times than they are now; and he studied the arguments of the leading contemporary theorists of scientific method, people like John Herschel and William Whewell. He formulated his theory as he read these philosophers. Their works affected his thinking, and sensitised him to the potential philosophical impact of his own ideas. Scribbling in a notebook while in his early thirties, Darwin could barely contain his excitement at the promise of his nascent evolutionary views: 'Origin of man now proved.—Metaphysic must flourish.—He who understands baboon would do more towards metaphysics than Locke' (Notebook M, in Barrett et al. 1987: 539).

Darwin's reference to 'metaphysics' might mislead modern philosophical readers. These days, 'metaphysics' is usually used to label the philosophical study of such things as causation, space or time – it is the study of fundamental questions about the nature of the universe. Back then, 'metaphysics' referred primarily to the study of the mind. The notebook which Darwin labelled 'M' is largely dedicated to philosophical and psychological reflections on the emotions, mental illness, language, ethics, knowledge and such like.

Darwin aspired to the status of, as he put it, a 'philosophical naturalist' (Sloan 2003). This label is also liable to mislead. As Darwin understood the phrase, it did not mean a naturalist who is interested in philosophy, but a naturalist who seeks a scientific explanation for the patterns observed in nature. A philosophical naturalist would not be content merely to describe and catalogue the species that populate the Earth, but would feel it necessary to say why there should be just those species, with just those proper-ties, rather than some other set of species, differently arranged. Darwin answered these questions by appealing to two ideas, not one. He argued that different species – our own included – are descended from common ancestors to form a great 'Tree of Life'. This is the hypothesis of evolution. Darwin also argued that natural selection was the agent primarily responsible for the shape of this tree. This combination of evolution and natural selection is what makes Darwin's natural history 'philosophical'.

We will see that Darwin was also a 'philosophical naturalist' in the historically incorrect sense of that phrase. He made efforts to relate what he referred to as 'my theory' to questions regarding politics, ethics and psychology. His notebooks feature frequent philosophical speculations and his published writings, which are far more cautious, nonetheless show considerable philosophical sensibility and engagement. We can see this sensibility manifested in Darwin's desire to ensure that the *Origin of Species* was structured in such a way that its argument met the highest standards of evidential rigour. We see it more directly in Darwin's choices of subject matter. Although the *Origin* remains largely silent regarding our own species, and merely hints at the promise Darwin sees in his view of life, ethics and politics come to the foreground in *The Descent of Man*. Here Darwin puts forward an evolutionary explanation of our ability to sense the difference between right and wrong, he suggests ways in which his natural historical approach recommends revisions to the abstract pronouncements of moral philosophers, and he considers the likely social consequences of the contemporary selection regime to which Victorian humans were subjected. As its title indicates, *Descent*'s successor volume – *The Expression of the Emotions in Man and Animals* – shines an evolutionary light on human and animal minds.

While Darwin rarely suggests that his natural historical reflections should wholly replace philosophical approaches to ethics, or the emotions, he does believe that philosophy is blind unless it is guided by evolutionary insights. As he explains in *Notebook N* (M's successor):

> To study Metaphysic, as they have always been studied, appears to me to be like puzzling at Astronomy without Mechanics.— Experience shows the problem of the mind cannot be solved by attacking the citadel itself.—the mind is a function of body.—we must bring some *stable* foundation to argue from.—
>
> (*Notebook N*, quoted in Barrett *et al.* 1987: 564)

Darwin's hope is that an evolutionary perspective – the perspective that recognises that species are modified versions of common ancestors – will provide us with some fixed points that can anchor and discipline philosophical speculations regarding human nature, and the human condition. These hopes are widely shared today. In 2006, Darwin's face is on £10 notes and evolution is everywhere. The past thirty years have seen an explosion of work applying evolutionary thinking to the emotions, ethics, culture, knowledge and many other topics that have traditionally fallen within the domain of philosophy. This book examines this evolutionary philosophical work.

2. DARWIN AND DARWINISM

A philosophical introduction to Darwin must be *philosophical*. My goal is not to explain how Darwin formulated his ideas, nor is it to set those ideas in the context of the time, nor is it to study how those ideas were received by Darwin's contemporaries. These are all interesting and important projects, but they are not mine. My goal is to examine the light which Darwin's ideas can shed on topics of philosophical importance. This means asking, for example, what difference a knowledge of our evolutionary history should make to how we understand human nature. Questions like these are evaluative: in blunt terms, we need to ask not merely how Darwin proposed that his ideas might influence philosophy, but whether these ideas can carry the philosophical weight that Darwin and others have placed on them.

This way of expressing things suggests a very sharp division between the job of the historian and the job of the philosopher. The philosopher asks whether Darwin's ideas (and the ideas of those who have built philosophically on Darwin's work) are any good; the historian asks how those ideas were formulated and received. In reality, the project of historical interpretation and the project of philosophical evaluation are not entirely independent of each other. Good philosophy is not the same as good history, but when philosophy is historically naive, it is likely to miss many opportunities. The risk that historical ignorance holds for philosophical evaluation is not so much that one ends up assessing a view that is not truly Darwin's (although one may well do that), it is that one ends up assessing a view that is unnecessarily weak. It is a good rule of philosophical method that if one is to investigate, for example, the linkage between evolution and ethics, one should pick the strongest, most plausible set of views for evaluation, rather than ideas that are obviously absurd or indefensible. The philosopher risks working with an inferior caricature of the views of those who are now dead if he or she is unable to interpret their words correctly. Accurate interpretation requires historical knowledge and historical sympathy.

Although this is a philosophical book about Darwin, I will not be claiming that Darwin was a philosopher on a par with the likes of Hume or Aristotle. He was, to repeat, a natural historian. Yet Darwin's philosophical interests and sensibilities led him to explore the broader significance of the evolutionary view of life. Darwin's writings are not unalloyed works of science that have subsequently been put to work by philosophers. Rather, Darwin presented his scientific ideas in a philosophically engaged manner, in a way which demanded, and continues to demand, further philosophical elaboration and exploration.

Darwin's ongoing influence can be measured by the extremely unusual role he plays among modern scientists. Many biologists have read Darwin's works. The works of Einstein, by contrast, although undoubtedly of enormous importance for modern physics, are rarely read by physicists working today. Modern biologists often refer to themselves as 'Darwinians'; one does not hear modern physicists describe themselves as 'Einsteinians'. When biologists differ over issues in modern science, they often try to claim Darwin for their team. Darwin is still regarded as a quotable biological authority, and struggles go on between biologists over how his views should be interpreted. Darwin is still a part of modern Darwinian biology in a way that Einstein is not a part of modern physics.

The story that will be told here is not, however, one of increasingly sophisticated philosophical engagement with an evolutionary theory that has remained static. Darwin's ongoing role as an authority figure might suggest that as far as science is concerned the basics of evolutionary biology have remained more or less the same since 1859, when the *Origin of Species* was published. In fact, we will see that there are significant differences between Darwin's views and those of modern evolutionary biologists. Darwin must consequently be distinguished from modern Darwinism. One of the primary justifications for examining Darwin's own views is precisely to expose the frequent mismatches between the Darwin who is invoked by today's biologists eager to defend their corner, and the Darwin who wrote the *Origin of Species* and the *Descent of Man*. Yet in spite of these differences, modern Darwinians regularly apply the concepts of modern evolutionary theory to the same issues – human nature, politics, the mind, knowledge, ethical judgement – that Darwin did. Our philosophical approach demands that in addition to examining Darwin's own writings on these issues, we ask whether more recent work has offered refinements or correctives to his arguments. This book does not aim to cover every philosophical problem in modern biology. Many of these modern problems, such as conceptual issues about the nature of genes, and their role in inheritance, can hardly be said to feature in Darwin's own work. But we will take account of modern evolutionary views when they relate to philosophical topics that Darwin discusses directly. This book is, therefore, a philosophical introduction to Darwinism, as well as a philosophical introduction to Darwin.

One might worry that this effort to compare modern views with those of a dead Victorian turns the book into an instance of so-called 'Whig' history of science – the kind of progressive history routinely maligned these days, which focuses selectively on those elements of the past that are important from the perspective of today's best science, as though the past were an engine for producing the textbooks of the present. This book is shamelessly evaluative, but many of its arguments are the opposite of Whiggish; perhaps they count as 'Tory' history of science. In some cases I will argue that Darwin's views are considerably more subtle, and more persuasive, than the stances adopted by modern-day philosophical naturalists. Darwin's philosophical views are of more than historical interest.

3. DARWIN UNFOLDING

A philosophical introduction to Darwin must be introductory. Philosophy strives for rigour and clarity, and it is inevitable that these standards must be compromised in a book that tries to cover as much ground as this one. Even inattentive readers will notice plenty of dotless 'i's and uncrossed 't's, and so I have given guides to further reading at the end of each chapter which list likely sources of dots and crosses.

Here is how the book is organised. The first chapter takes a brief look at Darwin's life. The remaining chapters are organised by topic. I have endeavoured to say something informative, and with luck something challenging and interesting, too, about the relevance of Darwin's work for the study of the mind, ethics, knowledge, and politics. These four topics are addressed in the core of the book – chapters five to eight respectively. But Darwin's discussion of these topics, and mine, too, relies on a prior understanding of the two ideas that lie at the foundation of his theorising. First is natural selection, which I discuss in chapter two. Second is Darwin's conception of species as genealogically related to form a giant family tree. This view is discussed in chapter three. Chapter four forms a kind of bridge between the early biological chapters and the later philosophical ones. It addresses Darwin's views about what makes a scientific theory a good one, and it is in this chapter that the relationship of Darwin's theory to various creationist views – including modern 'Intelligent Design Theory' – is discussed. The very last chapter concludes with some general reflections on Darwin's impact on philosophy as a whole. There is no one chapter that addresses the relevance of evolution for our conception of human nature. That theme runs right through the book, as it runs right through Darwin's work.

FURTHER READING

Readers seeking additional introductory material might turn to this recent collection of essays by leading philosophers and historians of biology. The contributions focus on, among other things, the formation of Darwin's theory, the relationship between Darwin's thinking and religion, and the philosophical influence of Darwin's work:

Hodge, J. and Radick, G. (eds) (2003) *The Cambridge Companion to Darwin*, Cambridge: Cambridge University Press.

A comprehensive single-author overview of Darwin's work and achievement, and a book whose ambitions are similar to this one, is:

Ghiselin, M. (1969) *The Triumph of the Darwinian Method*, Berkeley, CA: University of California Press.

One

Life

The philosophical spirit of broad-ranging and ambitious enquiry had a strong tradition in the Darwin family. Charles's paternal grandfather, Erasmus Darwin (1731–1802), was not only a successful medical doctor and investor, but also a member of the celebrated 'Lunar Society', a group of engineers, manufacturers, philosophers and others participant in the enlightenment project of improvement and investigation. The society included James Watt, Joseph Priestley and Josiah Wedgwood. Erasmus was the author of poems, works of science and commentaries on cultural progress: his poetic treatises on technological advancement usually rolled all three genres into one. His book *Zoonomia* defended an early evolutionary theory according to which all of plant and animal life originated from primitive 'filaments', endowed with a tendency to self-improvement over time. Erasmus' theory has little in common with the evolutionary views later defended by his grandson, and more in common with those of the French naturalist Jean-Baptiste de Lamarck (1744–1829), whose ideas we will meet in a moment. Charles came to distance himself from the ideas of both men. While he may have been drawn to daring theorising by works such as *Zoonomia*, he eschewed his grandfather's scientific method on the grounds that it lacked empirical discipline. Recalling his student days he says:

> I had previously read the *Zoonomia* of my grandfather, in which similar views [to Lamarck's] are maintained, but without producing

any effect on me. Nevertheless it is probable that the hearing rather early in life such views maintained and praised may have favoured my upholding them under a different form in my *Origin of Species*. At this time I admired greatly the *Zoonomia*; but on reading it a second time after an interval of ten or fifteen years, I was much disappointed, the proportion of speculation being so large to the facts given.

(*Autobiography*: 24)

Two other currents that ran strong in the Darwin family were medicine and money. Charles's father, Robert Waring Darwin (born in 1766), was a physician like Erasmus. He inspired great confidence in his patients, and his practice enjoyed success as a result. But the bulk of Robert Darwin's income came not from medicine, but from stocks, bonds, rents and mortgages. He had interests in roads, canals, agricultural land and a large part of the Wedgwood china factory. (His weight was formidable, as well as his bank balance; Charles remembered him as 'very corpulent . . . the largest man whom I ever saw' [ibid.: 11]). Politically, Robert Darwin was a Whig, strongly anti-Tory, a believer in industry and progress, a materialist, probably an atheist and a critic of aristocratic privilege. Even so, as Charles's biographer Janet Browne explains, he was no revolutionary: 'He put his faith in the idea of reform through legislation, and strong private opinions did not stop him encouraging professional relations with local Tory peers and churchgoing squires' (Browne 2003a: 9).

Charles Darwin was born in Shrewsbury on the 12th February 1809. Not until the beginning of his time at university did the good fortune of his birth dawn on him. As he recalled much later:

I became convinced from various small circumstances that my father would leave me property enough to subsist on with some comfort, though I never imagined that I should be so rich a man as I am; but my belief was sufficient to check any strenuous effort to learn medicine.

(*Autobiography*: 22)

He had surmised, quite correctly, that he would never need to earn a living. There was no financial imperative that he should follow in his father's professional footsteps.

Charles's mother, Susannah Darwin, was the daughter of Josiah Wedgwood, Erasmus's Lunar colleague and founder of the famous pottery at Etruria. She died in July 1817 when Charles was only eight years old. He remembered 'hardly anything about her except her death-bed, her black velvet gown, and her curiously constructed work-table' (ibid.: 6).

When his mother died Charles had already spent a few months at a local school run by a Unitarian minister the Reverend Case, and at the age of nine he began at Shrewsbury School, where, although very close to home, he boarded. The headmaster was one Samuel Butler, and Darwin was not impressed by his pedagogy:

> Nothing could have been worse for the development of my mind than Dr Butler's school, as it was strictly classical, nothing else being taught except a little ancient geography and history. The school as a means of education to me was simply a blank . . . .
>
> (Ibid.: 10)

Charles himself was no child prodigy:

> When I left the school I was for my age neither high nor low in it; and I believe that I was considered by all my masters and by my Father as a very ordinary boy, rather below the common standard in intellect. To my deep mortification my father once said to me, 'You care for nothing but shooting, dogs, and rat-catching, and you will be a disgrace to yourself and all your family.'
>
> (Ibid.)

This love of country sports brought Darwin into the open air and gave him contact with nature, but when he left Shrewsbury School in 1825 there was little to suggest that he would become a naturalist. He tells us rather impiously that his main interest at this time was killing things:

In the latter part of my school life I became passionately fond of shooting, and I do not believe that anyone could have shown more zeal for the most holy cause than I did for shooting birds. How well I remember killing my first snipe, and my excitement was so great that I had much difficulty in reloading my gun from the trembling of my hands. This taste long continued and I became a very good shot.

(Ibid.: 21)

2. FROM SPORT TO SCIENCE

Darwin first sought to follow the family tradition by becoming a doctor, and tradition dictated he should train in Edinburgh, where he duly began his medical studies in October 1825 at the age of sixteen. His heart was not in it: 'Dr Duncan's lectures on Materia Medica at 8 o'clock on a winter's morning are something fearful to remember. Dr Munro made his lectures on human anatomy as dull, as he was himself, and the subject disgusted me' (ibid.: 22). Darwin was squeamish. His disgust was provoked by dissection, and surgery was also an intolerable strain to the sympathies he felt for its victims:

I also attended on two occasions the operating theatre in the hospital at Edinburgh, and saw two very bad operations, one on a child, but I rushed away before they were complete. Nor did I ever attend again, for hardly any inducement would have been strong enough to make me do so; this being long before the blessed days of chloroform. The two cases fairly haunted me for many a long year.

(Ibid.: 23)

Walking, riding, botanising and zoologising around Edinburgh offered a welcome escape as he turned away from the trials of medicine, but still Darwin was a long way from travelling the straight road to becoming a man of science. Even geology, the subject in which he would first make his name, held little interest for him:

> During my second year at Edinburgh I attended Jameson's lec-
> tures on Geology and Zoology, but they were incredibly dull. The
> sole effect they produced on me was the determination never as
> long as I lived to read a book on Geology or in any way to study the
> science.
>
> (Ibid.: 25–26)

Edinburgh was not all bad. Darwin became a member of the
Plinian Society, a small student-run group which 'met in an
underground room in the university for the sake of reading papers
on natural science and discussing them' (ibid.: 24). One of the
most important friendships Darwin made there was with Robert
Grant, a lecturer at Edinburgh, and a follower of Lamarck. Grant
had been exposed to Lamarck's views during the time he had
spent in Paris, and Darwin remembered that: 'He one day, when
we were walking together burst forth in high admiration of
Lamarck and his views on evolution. I listened in silent astonishment, and as far as I can judge, without any effect on my mind'
(ibid.).

Darwin's astonishment at this praise for Lamarck owes itself to
the poor light in which Lamarck's theory was viewed around that
time, primarily in Britain, but also in France. Lamarck's ideas,
which were largely ignored until much later in the century, were
dismissed as speculative, with little evidential basis. He had argued
in favour of the potentially unlimited mutability of species over
time – the view known in France as *transformisme*, and in Britain as
transmutationism.

The caricature of Lamarck's position that we have inherited
today tends to stress two further themes: first is the view that species
adapt to their environments by conscious willing; second is the
so-called 'inheritance of acquired characteristics'. The overall
Lamarckian package has it that, for example, giraffes 'will' themselves to reach leaves at the tops of trees, and as a result their
necks get longer. Their offspring are then born with longer necks.
This is, as I say, a distortion of Lamarck's view (Bowler 1984:
81). He believed that as changing environments impose new
requirements on species, the organisms in question are forced to

acquire new habits in response to these altered demands. Conscious willing was not a part of Lamarck's theory; the view that it was can be traced to a hostile discussion of Lamarck by Charles Lyell, a British geologist who influenced Darwin deeply. Lamarck held that organisms have a kind of inner drive to adapt to their conditions. Lamarck did indeed believe that if a faculty was used during the life of an individual, then the alterations brought about by increased use would be inherited in future generations. So Lamarck held that adaptable habits, coupled with this mechanism of inheritance (called 'use-inheritance' in Britain), could lead to the limitless transformation of species in such a way that they would track the demands of their environments as those environments changed over time.

As this book unfolds we will see that Darwin, like Lamarck, came to believe that species were mutable, and he also believed (as did most of his contemporaries) in the importance of use-inheritance. Even so, there is one particular feature of Lamarck's view that is worth stressing, for it is quite different to any view Darwin came to hold. Today, when we think of 'evolution', we probably think of evolutionary trees that depict genealogical relationships between species, and which show how today's species are descended from a small number of common ancestors. That is an image of evolution that we owe primarily to Darwin. Lamarck, in contrast, believed that the simplest forms of life had been (and were still being) generated spontaneously from the coming-together of inert matter. According to Lamarck, once a simple life form is generated its descendants then undergo a series of transformations to become more complex as time goes by. But the species we see today do not have ancestors in common in the way that a Darwinian tree demands; instead, species' differing degrees of complexity reflect (as Lamarck saw it) the different periods of time that have elapsed since the spontaneous formation of those species' unrelated ancestors. If we find an organism today that is quite simple, this indicates that it must be descended from a spontaneously generated ancestor that appeared recently. More complex organisms are the descendants of organisms that were spontaneously formed far longer ago (ibid.: 79–81). Other important differences between Lamarck and Darwin will become apparent in later chapters, specifically regarding the mechanism by which organisms become adapted to their environments; for now it is important to note that they are also at odds over the historical pattern of evolutionary change.

By the end of Darwin's second year in Edinburgh it was clear that medicine was not for him. Robert encouraged his son to follow a career in the clergy, and although Charles had some doubts they were not enough to scupper the plan:

> I asked him to consider, as from what little I had heard and thought on the subject I had scruples about declaring my belief in all the dogmas of the Church of England; though otherwise I liked the thought of becoming a country clergyman.
>
> (*Autobiography*: 29)

It was a requirement that any Anglican priest should have a degree from one of the English universities, and so Darwin arrived in Cambridge, in January 1828, as a student of Christ's College. Darwin's comments on his Cambridge experiences make for mixed advertising. He claimed to have profited from its formal instruction no more than he did from that of Dr Butler: 'During the three years which I spent at Cambridge my time was wasted, as far as the academical studies were concerned, as completely as at Edinburgh and at school' (ibid.: 30). But, as at Edinburgh, there were good experiences as well as bad, including some of an intellectual kind. He profited especially from the works of William Paley. Paley had been a fellow of Christ's, and he is now best known for a canonical defence of the 'Argument from Design' in his lively book *Natural Theology*. This is the argument which uses the elegant design of organic nature – the eye that is so well-suited to sight, the wing to flight – as evidence for the existence of an intelligent creator competent to produce such workmanship. In later years Darwin concluded that his own views dealt the death blow to the design argument. At Cambridge, however, Darwin explains that:

> In order to pass the B.A. examination, it was, also, necessary to get up Paley's *Evidences of Christianity*, and his *Moral Philosophy*.

> This was done in a thorough manner, and I am convinced that I could have written out the whole of his *Evidences* with perfect correctness, but not of course in the clear language of Paley. The logic of this book and as I may add of his *Natural Theology* gave me as much delight as did Euclid. The careful study of these works . . . was the only part of the Academical Course which was of the least use to me in the education of my mind. I did not at that time trouble myself with Paley's premises; and taking these on trust I was charmed and convinced by the long line of argumentation.
>
> (Ibid.: 30–31)

Natural Theology was not, in fact, one of the texts on which Darwin was examined, but Darwin gained a good enough knowledge of Paley's other works to pass his ordinary degree (the BA) and gain 'a good place among the *hoi polloi*, or crowd of men who do not go in for honours' (ibid.).

The contacts which Darwin made at Cambridge were vital contributors to his subsequent successes. Edinburgh's unpleasant geological experiences had so turned him off the subject that he still could not bring himself to attend geology lectures (given by Adam Sedgwick). But he did befriend John Henslow, a young Anglican priest and Professor of Botany. Darwin attended Henslow's lectures, he accompanied him on long botanising walks, and frequently went to his house for dinner. This friendship, he later wrote, 'influenced my whole career more than any other' (ibid.: 34). At these dinners, Darwin met William Whewell (pronounced to rhyme with 'jewel'), one of the leading intellectual lights of Victorian society, an influential writer on scientific method, and a man who would eventually become Master of Trinity College:

> Dr Whewell was one of the older and distinguished men who sometimes visited Henslow, and on several occasions I walked home with him at night. Next to Sir J. Mackintosh he was the best converser on grave subjects to whom I ever listened.
>
> (Ibid.: 35)

Compared with his negative assessment of Cambridge's impact on his intellect, Darwin's verdict on the non-academic part of his life there is more likely to be used as a soundbite by University fundraisers: 'Upon the whole the three years which I spent at Cambridge were the most joyful in my happy life; for I was then in excellent health, and almost always in high spirits' (ibid.: 36). He developed a taste for painting ('that of Sebastian del Piombo exciting in me a sense of sublimity' [ibid.]) and music, although he was 'utterly destitute of an ear' (ibid.). Sport remained a significant interest, as well as the predictable pleasures of student life:

> Although . . . there were some redeeming features in my life at Cambridge, my time was sadly wasted there and worse than wasted. From my passion for shooting and for hunting and when this failed for riding across country I got into a sporting set, including some dissipated low-minded young men.
>
> (Ibid.: 31)

But in spite of this, natural history, and more specifically entomology, was beginning to command his attention: '. . . no pursuit at Cambridge was followed with nearly so much eagerness or gave me so much pleasure as collecting beetles' (ibid.: 32).

Darwin got his degree at the beginning of 1831, but University regulations demanded that he remain in residence in Cambridge for two more terms. He must have got over his loathing of geology during this period, for in August 1831, at Henslow's suggestion, Adam Sedgwick allowed Darwin to accompany him on a geological expedition to North Wales. This was clearly a stimulating trip, confirming Darwin's excitement at the possibilities offered by science and instructing him in method in the field. Even then, the conversion from sportsman to scientist was not complete. He left Sedgwick before the end of the expedition, going first to Barmouth and then to Maer (his uncle's home) 'for shooting; for at that time I should have thought myself mad to give up the first days of partridge shooting for geology or any other science' (ibid.: 36–37).

When he got back to Shrewsbury he found a letter from Henslow telling him of a Captain FitzRoy, who was willing to share his own cabin with 'any young man who would volunteer to go with him without pay as naturalist to the Voyage of the *Beagle*.' Robert Darwin was initially opposed to the idea (a sea-voyage was dangerous, and it would disrupt the progression of his son's clerical career), but Charles's uncle Jo (Josiah Wedgwood II) persuaded Robert to consent. A few days later Darwin went to Cambridge to see Henslow, then to London for an interview with FitzRoy. The matter was settled: Darwin made a brief visit to Plymouth on 11th September 1831, he went home again for a few weeks to say his farewells, and then returned to Plymouth on 24th October, to await the *Beagle*'s departure.

3. THE *BEAGLE* VOYAGE

The *Beagle* finally set sail on 27th December 1831, after two previous attempts to leave port were foiled by bad weather. It was a surveying ship, charged with charting the South American coast. The voyage was initially scheduled to last three years; in the end it lasted nearly five. Only twenty-two when he left England, it was over the first months of the voyage that Darwin turned decisively to science, coming to the realisation, 'though unconsciously and insensibly, that the pleasure of observing and reasoning was a much higher one than that of skill and sport' (ibid.: 43). He would regard this long adventure as the foundation of his successes as a naturalist, writing near the end of his life that:

> The voyage of the *Beagle* has been by far the most important event in my life and has determined my whole career . . . I have always felt that I owe to the voyage the first real training or education of my mind. I was led to attend closely to several branches of natural history, and thus my powers of observation were improved, though they were already fairly developed.
>
> (Ibid.: 42)

Aside from providing cultivated company to the ship's troubled captain Robert FitzRoy, Darwin's primary goal for the *Beagle* voyage was not to observe finches and giant tortoises, but to work on geology and invertebrates. He took with him the first of Charles Lyell's three-volume *Principles of Geology*, and the combination of close study of this work together with direct observation of varied geological phenomena soon converted Darwin to a Lyellian view of things. Unlike Sedgwick, Lyell was an advocate of *uniformitarianism*, the view that all geological phenomena could be explained by reference to the slow action of causes of the same type as we can observe acting around us now, compounded over vast stretches of time to produce such large scale features as mountains or oceans. This view was opposed to the *catastrophism* of Sedgwick and others – the view that the Earth's history had been punctuated by a series of swift and violent catastrophes of a character not experienced by men alive today.

Lyell was certainly no evolutionist: he had criticised Lamarck's transmutationism, arguing that species could vary, but were forever held within fixed boundaries. Even so, Darwin's conversion to uniformitarianism convinced him, in Janet Browne's words, 'that the majestic story of nature could be explained by the accumulation of little things' (Browne 2003a: 294). This gradualist theme would run through his career's work.

The story of the *Beagle* voyage is a long and fascinating one, and what is told here will focus only on highlights. The ship sailed south from Plymouth, arriving in the Cape Verde Islands (off the West African coast) in mid-January. From there they sailed to Bahia (Salvador) in Brazil, where Darwin experienced both slavery and a tropical rainforest for the first time. He argued with FitzRoy about slavery, a practice which Darwin 'abominated' (*Autobiography*: 40). The rainforest, on the other hand, was inspiring:

> Delight . . . is a weak term to express the feelings of a natu-
> ralist who, for the first time, has wandered by himself in a
> Brazilian forest. . . . To a person fond of natural history, such

a day as this, brings with it a deeper pleasure than he ever
can hope to experience again.

(Darwin 1913: 10–11)

The ship moved further down the coast, arriving in Rio de Janeiro
in April 1832, and Montevideo in July. It was at this time that the
Beagle's surveying business began, sailing down the east coast of
South America and up the west coast, not departing for the
Galapagos Islands until September 1835. Darwin made several
long trips inland during this time. Janet Browne paints a picture of
a young man revelling in a vigorous outdoor life, which poor
health would deny him almost as soon as he returned home. In a
letter to his sister Caroline he describes the sport in Patagonia:

In this line I never enjoyed anything so much as Ostrich hunting
with the wild soldiers, who are more than half Indian. They catch
them by throwing two balls which are attached to the ends of a
thong so as to entangle their legs: it was a fine animated chase.

(Quoted in Browne 2003a: 220)

One incident that perhaps coloured Darwin's view of race is worth
recounting. The *Beagle* was carrying three natives of Tierra del Fuego,
the group of islands at the southern point of South America. These
three (whom the British named Fuegia Basket, York Minster and
Jemmy Button) had been taken by FitzRoy and transported to
England on an earlier voyage. They had learned English, been
presented to the King and Queen, and their heads had been examined as the craze for phrenology – the study of character as revealed
in cranial bumps – demanded. FitzRoy was now returning them
home to Tierra del Fuego, where he intended to help found a
mission. The three newly civilised Fuegians would assist in its
running.

As the voyage drew on, Darwin warmed especially to Jemmy
Button:

Jemmy Button was a universal favourite, but likewise passionate;
the expression of his face at once showed his nice disposition.

He was merry and often laughed, and was remarkably sympathetic with any one in pain: when the water was rough, I was often a little sea-sick, and he used to come to me and say in a plaintive voice, 'Poor, poor fellow!' but the notion, after his aquatic life, of a man being sea-sick, was too ludicrous, and he was generally obliged to turn on one side to hide a smile or laugh, and then he would repeat his 'Poor, poor fellow!' He was of a patriotic disposition; and he liked to praise his own tribe and country, in which he truly said there were 'plenty of trees,' and he abused all the other tribes: he stoutly declared that there was no Devil in his land.

(Darwin 1913: 217–18)

Darwin's experiences with Jemmy did nothing to prepare him for the welcome which a band of unimproved Fuegians gave the *Beagle* as it neared land in December 1832:

I shall never forget how savage & wild one group was.—Four or five men suddenly appeared on a cliff near to us,—they were absolutely naked & with long streaming hair; springing forth from the ground & waving their arms around their heads, they sent forth most hideous yells. Their appearance was so strange, that it was scarcely like that of earthly inhabitants.

(From Darwin's *Beagle* diary, quoted in Browne 2003a: 240)

Darwin found it hard to imagine that Jemmy was of the same race as these Fuegians. The processes that had lifted Jemmy from such a low state suggested great possibilities for the bettering of man: 'What a scale of improvement is comprehended between the faculties of a Fuegian savage & a Sir Isaac Newton', Darwin wrote in his diary (quoted in ibid.: 248). It was therefore a shock to see that what goes up can also come down. The *Beagle* left Jemmy, York and Fuegia at the beginning of 1833. When it returned to Tierra del Fuego in March 1834, Darwin initially did not recognise Jemmy Button canoeing to meet them, apparently a 'savage' once more:

It was quite painful to behold him; thin, pale, & without a remnant of clothes, excepting a bit of blanket round his waist: his hair, hanging over his shoulders; & so ashamed of himself, he turned his back to the ship as the canoe approached. When he left us he was very fat, & so particular about his clothes, that he was always afraid of even dirtying his shoes; scarcely ever without gloves & his hair neatly cut.—I never saw so complete & grievous a change.

(Quoted in ibid.: 268–69)

Jemmy refused repeated offers to return to England with the *Beagle*. FitzRoy's mission station was abandoned.

Not until September 1835, four-fifths of the way through the *Beagle*'s voyage, did Darwin arrive in the Galapagos Islands. Today one sometimes has the impression that Darwin stepped off the boat, surveyed the different species of finch – all very similar, but each differently adapted to its own island – and intuited the truth of the theory of evolution by natural selection. This is not what happened. The principle of natural selection would not come to Darwin for several years, and it was not until he returned to London that he concluded that the different species of the Galapagos were formed by modification from common ancestors. When the local Vice-Governor told Darwin that tortoises of different forms were peculiar to different islands, he 'did not for some time pay sufficient attention to this statement, and I had already partially mingled together the collections from two of the islands' (Darwin 1913: 418). He classified what he later recognised as different species of finch in entirely different families, calling one a 'wren'. As he later put it: 'I never dreamed that islands, about fifty or sixty miles apart, and most of them in sight of one another, formed of precisely the same rocks, placed under a quite similar climate, would have been differently tenanted' (ibid.).

Darwin eventually noted that mockingbirds from two of the Galapagos islands tended to differ, and this made him look again at the birds he had seen earlier on a third island, James Island. He remarked in his field notes:

This bird which is so closely allied to the Thenca of Chili (Callandra of B. Ayres) is singular from existing as varieties or

distinct species in the different Isds.—I have four specimens from
as many Isds.—There will be found to be 2 or 3 varieties.—Each
variety is constant in its own Island.—This is a parallel fact to the
one mentioned about the Tortoises.

(Quoted in Browne 2003a: 304–5)

Darwin had realised that the different islands, although they had
almost identical environments, contained distinct species of bird.
He had also realised that these birds were similar to species found
in the nearest area of continental mainland. Logically speaking,
these observations point strongly to transmutation: the similarities
between these island species and those of the mainland can be
explained by the hypothesis that they are modified forms of an
earlier ancestor which existed in that geographical region.
Arguments like this would eventually appear in the *Origin of Species*.
But Darwin did not, in fact, become a convert to transmutationism until well after the *Beagle* left the Galapagos.

The *Beagle* sailed next to Tahiti, where a sober Darwin thought it
unbecoming that the Tahitian women should wear little save for a
flower in their hair. In December 1835 the ship arrived in New
Zealand. Perhaps through fatigue at a voyage that had lasted for four
years, and lacking any priming from Peter Jackson's films, Darwin
wrote:

> I believe we were all glad to leave New Zealand; it is not a
> pleasant place. Amongst the natives there is absent that
> charming simplicity which is found in Tahiti; and the greater part
> of the English are the very refuse of society. Neither is the
> country itself attractive.
>
> (Darwin 1913: 456–57)

He thought much better of Australia and prosperous Sydney,
which he reached in January 1836: 'My first feeling was to
congratulate myself that I was born an Englishman' (ibid.: 459).
In the spring they crossed the Indian Ocean by way of the Keeling
Islands and Mauritius, arriving at Cape Town at the end of May. It
was around this period – as late as this – that Darwin seems to have

abandoned the prospect of a country parsonage. In Cape Town Darwin dined with the astronomer Sir John Herschel, who was there to make various observations, including watching Halley's Comet. Darwin had a great respect for Herschel's philosophical works, perhaps because Herschel was of one mind regarding scientific method with Charles Lyell:

> I felt a high reverence for Sir J. Herschel, and was delighted to dine with him at his charming house at the C. of Good Hope and afterwards at his London house. I saw him, also, on a few other occasions. He never talked much, but every word which he uttered was worth listening to.
>
> (*Autobiography*: 62)

In July, when the *Beagle* arrived at Ascension Island, Darwin received news telling him that Henslow had edited some of his letters on scientific topics, which had then been read to considerable acclaim in London. Darwin was, it turned out, already well known to the minds whose thoughts he cared about: even Lyell, whom he had never met, looked forward to picking his brain.

From Ascension the Beagle returned once more to Bahia in Brazil, then travelled north to the Azores and the final run of the voyage. Darwin never found his sea-legs, even at the end of the journey. He was ill, off and on, over the whole five years, writing to his family: 'I loathe, I abhor the sea and all ships which sail on it. Not even the thrill of geology makes up for the misery and vexation of spirit that comes with sea-sickness' (quoted in Browne 2003a: 178).

4. LONDON, MARRIAGE AND THE NOTEBOOKS

The *Beagle* arrived in Falmouth on 2nd October 1836. Darwin initially went back to Shrewsbury, where his father greeted him by insinuating that the learning he had acquired in his five-year absence had caused his head to change shape. Charles decided that London was the place to build a scientific career and so moved there in March 1837, taking lodgings at Great Marlborough Street.

The period spent in London is now viewed as one of the most important for the formation of all Darwin's later evolutionary views (Hodge 2003). He gave numerous talks to the Geological Society, and began to publish regularly. One of his first publications was a personal narrative of his round-the-world trip entitled *Journal of Researches*, now more usually known under the title *The Voyage of the Beagle*. It was also in London that he transformed the *Beagle*'s mass of observations and experiences on matters geological, botanical, geographical, zoological, philosophical, anthropological and embryological into a coherent picture of life's unfolding. His reading, which was extensive at this time, included works of the philosophers David Hume and Adam Smith, as well as philosophical books less well known today such as Herbert Mayo's *Philosophy of Living*, John Abercrombie's *Inquiries Concerning the Intellectual Powers and the Investigation of Truth* and James Mackintosh's *Dissertation on the Progress of Ethical Philosophy*.

It was during the London period that Darwin began to speculate about the possibility of a transmutationist view of natural history, which he did in notebooks, opening one dedicated exclusively to an assessment of evidence for such a view a few months after arriving in the capital. Eight years later he had completed five books marked with letters 'A' to 'E', as well as the 'M' and 'N' notebooks, on metaphysics.

Darwin would later claim that transmutationist views had appeared plausible to him during his *Beagle* voyage. He had been, he said:

> Deeply impressed . . . by the South American character of most of the productions of the Galapagos archipelago, and more especially by the manner in which they differ slightly on each island of the group; none of these islands appearing to be very ancient in a geological sense . . . It was evident that such facts as these, as well as many others, could be explained on the supposition that species gradually became modified; and the subject haunted me.
>
> (*Autobiography*: 71–72)

Even so, he was not satisfied by a case of this sort, for it gave no indication of how species became so wonderfully adapted to their environments. Darwin could not account for the phenomena which Paley had taught him to admire: 'I had always been much struck by such adaptations, and until these could be explained it seemed to me almost useless to endeavour to prove by indirect evidence that species have been modified' (ibid.). In a side-swipe at Lamarck, Darwin says he thought it 'evident that neither the action of the surrounding conditions, nor the will of the organisms (especially in the case of plants)' could explain this good design (ibid.: 72).

On 28th September 1838, Darwin began reading Thomas Malthus's *Essay on the Principle of Population*. Malthus argued that populations have a tendency to expand over time, which outstrips the increase in the supply of food. The result, unless population growth is voluntarily held in check, is scarcity of resources, famine and an inevitable pruning of the population. Darwin would come to regard the reading of this book, which he completed on 3rd October, as an extended 'Eureka!' moment in the formulation of his views:

> In October 1838, that is, fifteen months after I had begun my systematic enquiry, I happened to read for amusement 'Malthus on *Population*', and being well prepared to appreciate the struggle for existence which everywhere goes on from long-continued observation of the habits of animals and plants, it at once struck me that under these circumstances favourable variations would tend to be preserved, and unfavourable ones to be destroyed. The result of this would be the formation of new species.
>
> (Ibid.: 72–73)

Although Darwin presents the principle of natural selection as largely worked out upon reading Malthus, he did not rush into print; indeed, until the publication of the *Origin of Species* some twenty years later he did not publish his theory of natural selection at all:

> Here, then, I had at last got a theory by which to work; but I was so anxious to avoid prejudice, that I determined not for some time

to write even the briefest sketch of it. In June 1842 I first allowed myself the satisfaction of writing a very brief abstract of my theory in pencil in 35 pages; and this was enlarged during the summer of 1844 into one of 230 pages, which I had fairly copied out and still possess.

(Ibid.)

Darwin's return from his travels also turned his mind to marriage, and here, too, he cautiously evaluated the pros and cons of such a transmutation in a series of notes. On the plus side, a wife promised 'female chit-chat', as well as someone 'who would feel interested in one'. A wife would be 'better than a dog anyhow'. (This is high praise – Darwin was extremely fond of dogs.) He expressed the negative points of marriage considerably more forcefully. A wife and family would be expensive, and they would consume time: 'Fatness & idleness—anxiety & responsibility—less money for books &c. If many children forced to gain one's bread'. Many doors would be closed to him, too: 'Eheu!! I should never know French,—or see the continent—or go to America, or go up in a Balloon, or take solitary trip in Wales—poor slave—you will be worse than a negro'. Even so, he remarks uncharacteristically that 'there is many a happy slave'. His romantic vision of a future as a ballooning bachelor needed bringing down to earth: 'Only picture to yourself a nice soft wife on the sofa with good fire, & books & music perhaps—Compare this vision with the dingy reality of Grt. Marlbro' St'. The matter was decided, 'Marry—Marry—Marry. QED' (all notes quoted in Browne 2003a: 379).

In the end it was Emma Wedgwood whom he targeted, the youngest daughter of his uncle Jo. They married on 29th January 1839, five days after Darwin was made a Fellow of the Royal Society, the elite association of men of science. Emma was genuinely fond of Charles, writing to her aunt that:

He is the most open, transparent man I ever saw, and every word expresses his real thoughts. He is particularly affectionate and very nice to his father and sisters, and perfectly sweet tempered, and possesses some minor qualities that add particularly to

one's happiness, such as not being fastidious, and being humane
to animals

(Quoted in ibid.: 393)

And although Darwin's initial decision to marry was calculated,
near the end of his life he wrote movingly of their relationship:

> She has been my greatest blessing, and I can declare that in my
> whole life I have never heard her utter one word which I had
> rather have been unsaid. She has never failed in the kindest sym-
> pathy towards me, and has borne with the utmost patience my
> frequent complaints from ill-health and discomfort. I do not
> believe she has ever missed an opportunity of doing a kind action
> to anyone near her. I marvel at my good fortune that she, so
> infinitely my superior in every single moral quality, consented to
> be my wife. She has been my wise adviser and cheerful com-
> forter throughout my life, which without her would have been
> during a very long period a miserable one from ill-health. She has
> earned the love and admiration of every soul near her.
>
> (*Autobiography*: 56)

Charles and Emma stayed in London at first, going to live on
Upper Gower Street, close to University College. His health
declined markedly during the London period, although there was
a temporary improvement in 1842 which allowed him to go once
again to Wales on a geological expedition, 'the last time I was ever
strong enough to climb mountains or to take long walks, such as
are necessary for geological work' (ibid.: 58). He remained strong
enough, however, 'to go into general society' (ibid.), where he
became well-acquainted with many of the leading men of science.
He saw Herschel again, and he got to know Charles Lyell, his
geological hero: 'I saw more of Lyell than of any other man both
before and after my marriage. His mind was characterised, as it
appeared to me, by clearness, caution, sound judgement and a
good deal of originality' (ibid.). He met others, too, such as
Herbert Spencer and Thomas Carlyle, neither of whom he had
much time for. Looking back he wrote that:

Herbert Spencer's conversation seemed to me very interesting, but I did not like him particularly, and did not feel that I could easily have become intimate with him. I think that he was extremely egoistical. After reading any of his books, I generally feel enthusiastic admiration for his transcendent talents, and have often wondered whether in the distant future he would rank with such great men as Descartes, Leibnitz, etc., about whom, however, I know very little.

(Ibid.: 63)

Darwin makes it clear that the absence of good empirical evidence left him unimpressed by much of Spencer's work:

His deductive manner of treating every subject is wholly opposed to my frame of mind. His conclusions never convince me: and over and over again I have said to myself, after reading one of his discussions,—'Here would be a fine subject for half-a-dozen years' work.'

(Ibid.: 64)

As always, Darwin continued to work, completing an important work on coral reefs soon after he and Emma were married. Their first child was born in December 1839, and Charles's ill health together with a growing family and the grime and noise of the city persuaded them to seek a home out of London. Charles and Emma hit on Down House in the Kent village of Down (now spelled 'Downe'), not far from London. They moved there in September 1842, and they stayed there for the rest of their lives.

5. DOWN ...

Although he was only thirty-three when he settled in Kent, Darwin's later autobiographical reminiscences (never intended for publication) effectively end with the move to Down.

During the first part of our residence we went a little into society, and received a few friends here; but my health almost always suffered

> from the excitement, violent shivering and vomiting attacks being
> thus brought on. I have therefore been compelled for many years
> to give up all dinner-parties; and this has been somewhat of a
> deprivation to me, as such parties always put me into high spirits.
> From the same cause I have been able to invite here very few
> scientific acquaintances.
>
> (*Autobiography*: 68)

His health would not permit him to do anything lively, and so he
has nothing of interest in his life to describe after 1842 except for
a formidable series of what academics now call 'research outputs':

> My chief enjoyment and sole employment throughout life has
> been scientific work; and the excitement from such work makes
> me for the time forget, or drives quite away, my daily discomfort.
> I have therefore nothing to record during the rest of my life,
> except the publication of my several books.
>
> (Ibid.: 69)

Thanks to Darwin's son Francis, we know of Charles's daily
routine in some detail. He woke early and went for a short walk
before breakfast, which he took alone at about seven forty-five.
His best working hours were between eight and half-past nine,
after which he read letters, and often had a portion of a novel read
aloud to him. Work would begin again at half-past ten, and
conclude at noon. According to Francis, 'By this time he considered his day's work over, and would often say, in a satisfied voice,
"I've done a good day's work"' (Darwin 1905: 91). Whatever the
weather, he would then go for another walk, often beginning by
checking on the experimental plants in his greenhouse. After
lunch he lay on the sofa reading the newspaper, and then
progressed to writing letters. At around three he took another rest,
and smoked a cigarette while listening to Emma reading from a
novel. Then at four he took another walk. He would put an hour's
work in between half-past four and half-past five, followed by
another rest and more reading from a novel. Then after dinner he
would play two games of backgammon with Emma, and read

from a scientific book, or listen to her playing the piano. Francis concludes his description of his father's daily life:

> He became much tired in the evenings, especially of late years, when he left the drawing-room about ten, going to bed at half-past ten. His nights were generally bad, and he often lay awake or sat up in bed for hours, suffering much discomfort. He was troubled at night by the activity of his thoughts, and would become exhausted by his mind working at some problem which he would willingly have dismissed.
>
> (Ibid.: 102)

Darwin saw few people, and whether his illness was always the cause of this or sometimes a pretext, his isolation enabled him to work exceptionally hard. Yet one should not infer from this that he was a lone genius, single-handedly revolutionising biology from the sequestered comfort of Down. Darwin's scientific insights were far from solo efforts. He was a prolific correspondent, sending letters to all parts of the world. These letters did not merely ask questions; they regularly sought to persuade others to conduct small experiments, make observations, or oversee surveys on his behalf.

As we have seen, Darwin wrote a substantial essay on natural selection in 1844, which he chose not to make public. His illness seems to have alarmed him so much that he arranged with Emma that she should have this sketch published if he should die. He was indeed very ill, so much so that when his father died in 1848 he was unable to attend the funeral in Shrewsbury. But if Darwin had put together the primary outline of his theory by 1844, why did he not go public with it until 1859, when he wrote the *Origin of Species*? Why did he instead spend eight years, from 1846 to 1854, working on a series of anatomical studies of barnacles?

The question of why Darwin delayed is of considerable interest to historians (e.g. R. Richards 1983). Rather than give a decisive answer here, let me instead canvas some of the rationales that have been put forward. First, there are scientific reasons. Darwin's barnacle work was important for helping him to appreciate the range of variation in nature, even among the most important anatomical structures. His knowledge of barnacles led to him making significant changes to the evolutionary theory described in the essay of 1844. Second, there are personal reasons. Emma was a traditional Anglican, unlike Charles who already had doubts about many of the doctrines of the Church of England. Charles was no atheist, but perhaps he did not wish to hurt her by a public proclamation of his views, which undermined literalist readings of the Bible and any view of God as the immediate creator of species. Robert Darwin had warned him against any expression of religious doubt to his spouse:

> Nothing is more remarkable than the spread of scepticism or rationalism during the latter half of my life. Before I was engaged to be married, my father advised me to conceal carefully my doubts, for he said that he had known extreme misery this caused with married persons.
>
> (*Autobiography*: 55)

Third, there are reasons associated with reputation and rhetoric. It is likely that Darwin was moved by the hostile reactions which transmutationist views were receiving from those whom Darwin admired most. Lyell, as we have seen, had dismantled Lamarck's theory piece by piece. Then, in October 1844, the book *Vestiges of the Natural History of Creation* was anonymously published, a popular success and a public sensation. This book also argued for a transmutationist position, albeit one far more ambitious, and far more speculative, than anything Darwin would ever avow. It covered not only the development of plants and animals, but of men and women, and of the universe itself. It was dismissed by Whewell, Herschel and Sedgwick. Sedgwick was particularly hostile, writing in a review that it showed 'the glitter of gold-leaf without the solidity of precious metal'. It was shoddy work, and irresponsible, too: Sedgwick wrote to Lyell that 'If the book be true, the labours of sober induction are in vain; religion is a lie; human law a mass of folly and a base injustice; morality is moonshine; our labours for the black people of Africa were works

of madmen; and men and women are only better beasts!'
(quoted in Browne 2003a: 468). If Darwin took his time, he
might wait for a calmer sea on which to sail his own transmutationist theory; he might amass enough sober evidence and
marshal it in such a way as to pre-empt any doubt of the purity of
his metal; and he might find a way to judiciously avoid discussion
of man and morality. Darwin would aim to produce a work of
pure gold that was not so brightly buffed as to draw too much
attention to its implications. Perhaps he should cut his teeth on
barnacles.

Not until 1854, when the barnacle project was done, did
Darwin begin to work on a defence of his transmutationist views.
He assembled notes, discussed his views with friends and
colleagues, performed small experiments, and finally decided in
May 1856 (on Lyell's advice) to write a book, which he planned
to call *Natural Selection*. Had he completed it, this book would have
been three or four times the size of the *Origin of Species*. But Darwin
was pushed into hasty publication of his theory in a wholly
different format by the arrival, in June 1858, of a letter from
Alfred Russel Wallace, a young naturalist who was working in the
Dutch East Indies. Wallace had sent him an essay which, Darwin
judged, 'contained exactly the same theory as mine' (*Autobiography*:
73). Wallace had hit on a principle that was similar to Darwin's
natural selection, and like Darwin he claimed inspiration from
Malthus. Lyell and Joseph Hooker (a botanist friend, who would
eventually become Director of the Royal Botanical Gardens at
Kew) persuaded Darwin that his views and Wallace's should be
presented jointly, at a scientific meeting which would take place
on 1st July at the Linnaean Society in London. Neither Darwin nor
Wallace was present at the meeting. Extracts from Darwin's 1844
essay were read first, then a letter he wrote in 1857 to Asa Gray
(Harvard's professor of Botany), and finally Wallace's essay of
February 1858. Wallace, who was still travelling, knew nothing of
the plotting of Darwin's friends, and had no opportunity to
express a view about their plan.

Once outed as a transmutationist, there was no sense in Darwin
hiding his views any longer, and everything to be gained from

publishing a substantial work that might cement his claim to priority over Wallace. In August of 1858 Darwin began composing what he described as an 'abstract' of his theory. The manuscript was completed in May 1859. John Murray agreed to publish the work, and once proof-reading was done the abstract appeared on 24th November 1859 under the title *On the Origin of Species by Means of Natural Selection, or the Preservation of Favoured Races in the Struggle for Life*. Unlike *Vestiges of Creation*, the work included nothing about the origins of life itself, nothing about the origins of the Universe, and nothing, save for a small hint, about man. Darwin limited himself to the promise that:

> In the distant future I see open fields for far more important researches. Psychology will be based on a new foundation, that of the necessary acquirement of each mental power and capacity by gradation. Light will be thrown on the origin of man and his history.
>
> (*Origin*: 458)

Darwin followed the *Origin* with a work on orchids, and a large book on variation in plants and animals, in which he sketched the theory of inheritance that had been missing from his earlier books. Public debate over evolution was fierce in the 1860s, and Darwin became famous as a result of it. Light was eventually thrown on our own species, but Darwin waited until 1871 to do it when he published, aged 62, *The Descent of Man*. The book was a commercial success, bringing a good profit to Darwin and to John Murray. As Janet Browne notes, most reviewers expressed considerable discomfort at the assertion that man was descended from animals, but they were respectful of the book's author. The debate over evolution had moved on from the vitriolic years immediately following the *Origin*'s publication.

6. . . . AND OUT

Darwin continued to work, to experiment, to correspond and to publish throughout the 1870s, producing works on climbing plants, on emotions in humans and animals, and his last book (in 1881) on earthworms. In 1876, in his late sixties, he began to put on paper some autobiographical remarks, primarily intended, it seems, for the edification and instruction of his children. In them, Darwin laments what seemed to him a slow decline into philistinism. Back in his twenties he had taken pleasure in poetry, music and art. None of these, and not even the contemplation of a landscape, brought him pleasure any more. Instead he 'blesses all novelists', and the escape they bring him. Having written so much of a scientific nature, he reflects sadly that:

> My mind seems to have become a machine for grinding general laws out of large collections of facts, but why this should have caused the atrophy of that part of the brain alone, on which the higher tastes depend, I cannot conceive.
>
> (*Autobiography*: 85)

In these last years he describes himself as an 'agnostic'; in earlier times he had been a believer. On the *Beagle* voyage he was 'quite orthodox', but his faith waned with time:

> In my Journal I wrote that whilst standing in the midst of the grandeur of the Brazilian forest, 'it is not possible to give an adequate idea of the higher feelings of wonder, admiration, and devotion which fill and elevate the mind.' I well remember my conviction that there is more in man than the mere breath of his body. But now the grandest scenes would not cause any such convictions and feelings to arise in my mind.
>
> (ibid.: 52–53)

Soon after returning from his voyage, Darwin became sceptical of such things as the truth of the Old Testament ('no more to be trusted than the sacred books of the Hindoos, or the beliefs of any barbarian' [ibid.]), and the New Testament miracles. As the years passed he gradually gave up on Christianity altogether, for simple lack of evidence in favour of its claims. In his *Autobiography*, he assesses this religion frankly and with bitterness:

Thus disbelief crept over me at a very slow rate, but was at last complete. The rate was so slow that I felt no distress, and have never since doubted even for a single second that my conclusion was correct. I can hardly see how anyone ought to wish Christianity to be true; for if so the plain language of the text seems to show that the men who do not believe, and this would include my Father, Brother and almost all my best friends, will be everlastingly punished.

And this is a damnable doctrine.

(Ibid.: 50)

In spite of his rejection of Christianity in the years between the *Beagle*'s return and the *Origin*'s publication, and in spite of his persistent view that 'Everything in nature is the result of fixed laws' (ibid.), Darwin recollects that he retained a belief in God over this period:

Another source of conviction in the existence of God, connected with the reason and not with the feelings, impresses me as having much more weight. This follows from the extreme difficulty or rather impossibility of conceiving this immense and wonderful universe, including man with his capacity of looking far backwards and far into futurity, as the result of blind chance or necessity. When thus reflecting I feel compelled to look to a First Cause having an intelligent mind in some degree analogous to that of man; and I deserve to be called a Theist.

This conclusion was strong in my mind about the time, as far as I can remember, when I wrote the *Origin of Species*; and it is since that time that it has gradually with many fluctuations become weaker.

(Ibid.: 53)

⸱ʰ⸱⸱ ⸱f the *Origin* was not an atheist; he was instead swayed
⸱o God as one who first sets the lawful Universe in
⸱fter the *Origin*'s publication he became sceptical of

this theism, asking (in a way reminiscent of the philosopher David Hume): ' . . . can the mind of man, which has, as I fully believe, been developed from a mind as low as that possessed by the lowest animal, be trusted when it draws such grand conclusions?' (ibid.: 54). He summarises his stance: 'The mystery of the beginning of all things is insoluble by us; and I for one must be content to remain an Agnostic' (ibid.). The tone of Darwin's discussion suggests that 'atheist' might be a better label after all, for he assaults every reason one might claim for belief in God, coming to rest on the view that it is useless to speculate on what, if anything, sets the great machine in motion.

Darwin died on the 19th April 1882. He was seventy-three years old. He is said to have whispered to Emma, 'I am not in the least afraid to die. Remember what a good wife you have been'. His scientific friends asked the president of the Royal Society to request that he be buried in Westminster Abbey, and the funeral was held there one week after his death.

SUMMARY

As a young man, Charles Darwin was primarily interested in field sports. He was no great scholar, and first set himself on a career in medicine, then the clergy. The *Beagle* voyage made him as a natural historian. It cemented his passion for science, and it provided him with a fund of observations relating to geology, botany, zoology, embryology, anthropology and other branches of learning, which he continued to draw upon for the rest of his life. Darwin's broad approach to natural history was influenced by the geologist Charles Lyell, and his evolutionary views owe a great deal to Lyell's belief in the slow accumulation of minor causes to produce major effects. The *Origin of Species* – Darwin's defence of transmutation – did not appear until Darwin was fifty years old, and already well-regarded in the scientific establishment. It said almost nothing about the topic of man, and although his notebooks are full of speculation regarding evolution and the human species, Darwin did not speak out on these issues until *The Descent of Man*. From his early thirties until his death at the age of seventy-three Darwin led a secluded but busy life at Down House in Kent. His chronic illness prevented him from frequent socialising, and he passed his time writing scientific works, performing experiments in his house and garden, and gleaning information and opinion from a vast number of correspondents spread around the world.

FURTHER READING

Darwin's own somewhat unreliable reminiscences (including the memories put onto paper in 1876, which are widely quoted in this chapter, as well a much earlier autobiographical fragment written in August 1838) are collected in:

Darwin, C. (2002) *Autobiographies*, M. Neve and S. Messenger (eds) London: Penguin Classics.

Darwin's account of the Beagle voyage is also widely available, and the Penguin edition has a useful introduction:

Darwin, C. (1989) *The Voyage of the Beagle*, J. Browne and M. Neve (eds) London: Penguin Classics.

There are two outstanding and comprehensive biographies of Darwin. This chapter draws very heavily on Janet Browne's two-volume masterpiece. The first volume (*Voyaging*) was originally published in 1995, the second (*The Power of Place*) was originally published in 2002. Readers are more likely to get hold of the new Pimlico edition:

Browne, J. (2003) *Charles Darwin*, London: Pimlico.

The other leading biography, which puts more stress on the topics of class and religion, is:

Desmond, A. and Moore, J. (1992) *Darwin*, London: Penguin.

For a detailed account of Darwin's experiences in the Galapagos, readers should turn to the work of historian Frank Sulloway. His work on Darwin's interpretation of the Galapagos finches is especially well-known:

Sulloway, F. (1982) 'Charles Darwin's Finches: The Evolution of a Legend', *Journal of the History of Biology*, 15: 1–53.

Two

Selection

I. EVOLUTION AND NATURAL SELECTION

Are the many species that populate the Earth individually created by the intervention of a supernatural power, or are they instead modified forms of earlier ancestors, produced by natural means? In Darwin's time, the first view was known as the hypothesis of *special creation*. The second view – what we now call the hypothesis of evolution – was more normally known as *transmutationism* or, in France, *transformisme*. Darwin was not the first to suggest, nor even to provide evidence for, evolution. Indeed, later editions of the *Origin* begin with a 'Historical Sketch', where Darwin briefly runs through some of evolution's earlier advocates, including the French naturalists Buffon, Lamarck and Geoffroy St-Hilaire. As we have seen, there were evolutionists in Britain too, such as Robert Chambers, the man who turned out to have written *Vestiges of Creation*, first published fifteen years before the *Origin*.

Buffon, writing in the eighteenth century, espoused a highly restricted form of evolution. He did not argue that all animals formed a genealogical tree, nor even all members of a particular class, such as the mammals, but merely that the members of a family (all species of cat, for example) were related to each other (Bowler 1984: 70). Geoffroy St-Hilaire's *transformisme* was put forward only tentatively. Writing in the early nineteenth century he noted that species in quite different environments (whales in water, bats in the air and dogs on the ground) nonetheless show striking anatomical similarities (in terms of the structure of their limbs, for example). He suggested that these facts could be explained by thinking of species as differently transformed versions of a common ancestor. In the Origin's opening pages Darwin agrees that evidence of this kind is highly suggestive of evolution, but, he says: 'such a conclusion, even if well founded, would be unsatisfactory, until it could be shown how the innumerable species inhabiting this world have been modified, so as to acquire that perfection of structure and coadaptation which most justly excites our admiration' (Origin: 66). He goes on to mock Vestiges:

> The author of the 'Vestiges of Creation' would, I presume, say that, after a certain unknown number of generations, some bird had given birth to a woodpecker, and some plant to the misseltoe, and that these had been produced perfect as we now see them; but this assumption seems to me to be no explanation, for it leaves the case of the coadaptations of organic beings to each other and to their physical conditions of life, untouched and unexplained.
>
> (Ibid.: 67)

Darwin is laying down a challenge for transmutationism: until evolutionists have some plausible hypothesis that can account for the good fit between each species and its environment, there remains an avenue of argument open to the special creationists. They can argue that a supernatural intelligence has seen to it that each species is well-suited to its surroundings. So any evidential advantage evolution gains over special creation from its ability to account for resemblances between species risks being neutralised by its corresponding disadvantage in accounting for adaptation. Darwin invokes natural selection to explain adaptation, thereby filling the hole in his predecessors' arguments for evolution.

   Our first major task in this book is to explore Darwin's argument for the efficacy of selection as the agent of adaptation. As we investigate this, we will generate answers to two further questions. One concerns the similarities and differences between natural selection as Darwin understood it and selection as it is understood today. The second concerns the similarities and differences between natural selection's 'design' of beneficial adaptations and a designer's crafting of useful gadgets.

2. THE ARGUMENT FOR NATURAL SELECTION

Here, in one of the most quoted passages in his oeuvre, Darwin gives a précis of the long argument that seeks to establish both the reality of natural selection, and its efficacy in producing a good fit between organism and environment:

> If, during the long course of ages and under varying conditions of life, organic beings vary at all in the several parts of their organisation, and I think this cannot be disputed; if there be, owing to the high geometrical powers of increase of each species, at some age, season or year, a severe struggle for life, and this certainly cannot be disputed; then, considering the infinite complexity of the relations of all organic beings to each other, and to their conditions of existence, causing an infinite diversity in structure, constitution, and habits, to be advantageous to them, I think it would be a most extraordinary fact if no variation ever had occurred useful to each being's own welfare, in the same way as so many variations have occurred useful to man. But if variations useful to any organic being do occur, assuredly individuals thus characterised will have the best chance of being preserved in the struggle for life; and from the strong principle of inheritance they will tend to produce offspring similarly characterised. This principle of preservation, I have called, for the sake of brevity, Natural Selection.
>
> (ibid.: 169–70)

Darwin's argument in this paragraph is presented in seven main steps, which I have numbered in the following presentation. He first requests that we grant him two premises, both of which he regards as undeniable. Individual organisms in a species vary (I), and there is in nature a 'severe struggle for life' (II). This notion of the struggle for life is one that Darwin takes from the political economist the Reverend Thomas Malthus, whose ideas we met very briefly in chapter one. Malthus had argued in his *Essay on the Principle of Population* (first published in 1789) that human populations have a tendency to increase in *geometrical* ratio (i.e. the number of individuals in a later generation is calculated by *multiplying* the number of individuals in the previous generation by some factor), while the supply of food increases only in an *arithmetical* ratio (i.e. the food supply in a later generation is calculated by *adding* some figure to the food supply of the previous generation). So, for example, while the population might increase over four successive generations following a ratio 2:4:8:16, the food supply might increase over the same period following a ratio 2:4:6:8. If Malthus's premises are accepted then we can deduce that human populations will inevitably grow to outstrip the food supply. Regular famines will bring the population back into line with what natural resources have to offer; the only way to avoid such famines is through restrictions on the tendency to reproduce. Darwin sees (as Malthus did, too) that Malthus's principles apply not just to humans, but to the animal and plant worlds, where the tendency to reproduce also runs ahead of the supply of required natural resources, resulting in a struggle for life between individuals of the same species (and not, as is sometimes thought, between different species).

Darwin then argues that if we grant his first two premises, we should also grant, as consequences, first that it is very likely that some of the variations we see between individuals within a species will promote their bearers' welfare (III), and second that beneficial variations of this kind will be favourable in the struggle for existence (IV). Earlier in the same chapter he presents the reader with the simplified example of a pack of wolves which prey on fast-running deer: 'I can under such circumstances see no reason to doubt that the swiftest and slimmest wolves would have the best chance of surviving, and so be preserved or selected ... ' (ibid.: 138). Darwin's argument continues: whichever organisms have variations favourable in the struggle for existence will, *ipso facto*, have greater chances of living longer, and consequently of leaving offspring, than do others without these beneficial variations (V). Finally, Darwin requests a further premise, the 'strong principle of inheritance', which says that offspring tend to resemble their parents (VI). If this principle holds good, then those variations which better equip organisms in the struggle for existence will also be the ones that are 'preserved' in the offspring generation (VII). Given enough time, Darwin believes, the preservation of favourable variations can result in the production of the most exquisite adaptations. The principle of natural selection is, he writes, 'the doctrine of Malthus, applied to the whole animal and vegetable kingdoms' (ibid.: 68).

Before moving on, we should briefly explain the important distinction Darwin draws between natural selection and what he calls 'sexual selection'. The basic idea behind sexual selection is intuitive enough. Individuals face a struggle to find a mate, which parallels the struggle for existence. Some will be well-suited to the mating marketplace, others less so, and those individuals best equipped will tend to see the characteristics that led to their amorous successes represented in greater proportions in future generations. There is no requirement that the traits promoted by sexual selection need be of practical value to the organism, for sexual selection can favour traits that conform to wholly whimsical or arbitrary mate preferences of the opposite sex. Darwin's view is that in many species the dominant form of sexual selection is driven by females' criteria for mate-choice, which in turn dictate the traits acquired by males. In sum, Darwin argues that sexual selection can produce traits that assist in securing mates – peacocks' tails being the best-known example – in roughly the same way that natural selection produces adaptations that assist in the struggle for life.

3. DARWIN AND LAMARCK

Many of Darwin's contemporaries were unimpressed by the principle of natural selection. Darwin learned that Herschel had called it 'the law of higgledy piggledy'. Yet his idea was seized enthusiastically by the leading American philosophers of the late nineteenth and early twentieth centuries, such as Charles Sanders Peirce, John Dewey and William James. In an insightful, albeit obscure, essay published two years before Darwin's death, James gives a good account of what made Darwin's explanation of adaptation novel. Before Darwin, James says, adaptation was always explained by a single-stage process, in which an organism responds directly to its environment:

> The exercise of the forge makes the right arm strong, the palm grows callous to the oar, the mountain air distends the chest, the chased fox grows cunning and the chased bird shy, the arctic cold stimulates the animal combustion, and so forth.
>
> (James 1880: 444)

This is indeed the case with Lamarck, who thought that organisms had a kind of internal drive to adapt to the demands of their surroundings. Darwin's explanation is quite different:

> Darwin's first achievement was to show the utter insignificance in amount of these changes produced by direct adaptation, the immensely greater mass of changes being produced by internal molecular accidents, of which we know nothing.
>
> (Ibid.)

In other words, Darwin's breakthrough is to explain adaptation through a two-stage process, which couples chance variation to an environment that preserves any variations that happen to be favourable. When Darwin says that variations arise by chance, he means simply that we are ignorant of what explains them, as they are 'due to causes far too intricate to be followed' (Descent: 207). Variation is not in itself adaptive; there is no internal adaptive drive of the kind Lamarck envisages.

One might complain that Darwin has taken our ignorance about adaptation and replaced it with ignorance about variation. But whenever science makes progress our attention is drawn to new things we do not know; the trick is to replace old problems with new ones. If adaptation is explained in Lamarck's way by positing a tendency within organisms to produce adaptive variation, then we are left with the same problem we started with – the problem of explaining how the state of being well-suited to the environment comes about. For how is the adaptive drive supposed to work? What explains the success of organisms in responding appropriately to changing environmental demands? If adaptation is explained in Darwin's way by the interplay between chance variation and the environment, then we have progressed to a new set of problems. As we will see in section five, Darwin still owes us an account of the general *character* of variation, but James sees rightly that Darwin's position gains from the fact that he can confess his ignorance of variation's *causes*.

4. 'DARWIN'S DANGEROUS IDEA'

The précis that Darwin gives of natural selection is fairly simple. What one finds at the beginning of modern evolution textbooks is often even simpler (e.g. Ridley 1996: 71–73). There it is quite usual to read that selection acts whenever three conditions are met: organisms must vary, they must differ in fitness, and offspring must have a tendency to resemble their parents. One often reads, in summary form, that selection acts whenever there is 'heritable variation in fitness'; however, some caution is needed with this definition, for 'heritability' has a variety of technical senses in biology which mean we should not equate a 'heritable' trait with a trait that is inherited. I will not discuss the technical senses of 'heritability' in this book: interested readers might begin to investigate them by consulting a classic article by Lewontin (1985).

There remains one term we have not yet addressed in this trio of 'heritable variation in fitness'. To say what 'fitness' means is no easy matter, and one cannot turn to Darwin for help. He was persuaded by Herbert Spencer to use the phrase 'the survival of the fittest' in later editions of the *Origin* as a synonym for natural selection, but Darwin did not use the abstract noun 'fitness' in the way we do now. Today, 'fitness' is defined in many ways, but usually it has something to do with reproductive output. For the moment, it will be enough to think of fitness as a property possessed by individual organisms, which reflects their chances of leaving offspring: to say that you have high fitness is simply to say that you are likely to have many offspring. There is, then, no straightforward conceptual link between having high fitness and being well 'fitted', or suited, to your environment.

The modern argument linking selection to adaptation is similar to Darwin's, but has fewer steps. Suppose, as Darwin does, that offspring resemble their parents. Suppose, also, that organisms vary. If some variation augments an individual's fitness (i.e. its chances of reproducing), then this kind of variation will tend to appear in greater proportions in the offspring generation than in the parental generation. (Biologists say that the variation increases its frequency.) Thus, fitness-enhancing variations are preserved. Consider Darwin's example: if running fast increases a wolf's chances of leaving offspring, and if offspring resemble their parents in terms of running speed, then the proportion of fast-running wolves is likely to be higher in the offspring generation than in the parental generation. This process can be repeated generation after generation, making wolves faster and faster runners.

The philosopher Daniel Dennett tends to describe selection as a process of 'generate-and-test' (Dennett 1995). Forms are generated and tested against an environment. Some of these forms will have a tendency to reproduce more effectively than others, in virtue of some feature or another. These features – what Dennett calls 'good tricks' – are thereby preferentially retained in subsequent generations and a new search begins for tricks that are better still. Improvements once found are not lost; rather, they spread and act as the foundation for further improvements. This simple algorithm of generate-and-test, or natural selection, ultimately explains such marvels as the eye. Small wonder Dennett calls natural selection 'Darwin's Dangerous Idea' (ibid.).

If such a simple idea can explain so much, it is hardly surprising that when he heard about Darwin's theory, Darwin's great ally Thomas Henry Huxley exclaimed 'How extremely stupid not to have thought of that!' (quoted in Browne 2003b: 92). Huxley's remark is frequently cited today to suggest that natural selection is one of those rare ideas whose enormous power is obvious to all straight-thinking people, but only after a genius like

Darwin has alerted them to it. This is, I believe, a misleading picture of the relationship between natural selection and adaptation. The picture also makes it hard to understand why so many of Darwin's contemporaries – Huxley included – accepted his claim that life had evolved, without accepting his claim that natural selection acting on gradual variation was the primary agent of evolutionary change. If the abstract conditions of heritable variation in fitness are enough to bring about complex adaptation, then the only way to understand scepticism about selection in the late nineteenth century is by representing Darwin's scientific peers as either insincere or stupid. One of my key aims in this chapter is to show that Darwin's idea is more complex than first meets the eye.

5. NATURAL SELECTION AND VARIATION

Let us begin with some fairly obvious gaps that need to be filled if we are to make a case for selection's ability to explain the emergence of complex adaptations like eyes or wings. Natural selection can only preserve the fittest variant that happens to be available in any given generation. So if selection is to do what Darwin claims it can do – if, that is, selection is to build complex adaptations – there needs to be a broad supply of variation available for it to act on. It is not enough for organisms to vary; variation needs to be plentiful.

A second problem that confronts selection's ability to explain adaptation arises from the fact that organisms are integrated wholes (Gould and Lewontin 1979). They are not bundles of isolated traits, any one of which can be modified independently of all the others; changes to one organ, for example, may have knock-on effects on other parts of the system. Dennett knows this – indeed he discusses the point explicitly – but his coupling of selection understood as 'generate-and-test' with his frequent use of analogies between natural selection and product engineering might lead the unwary reader to the view that natural selection builds an organism in precisely the same way that a designer produces a new kind of washing machine. If you are trying to invent a better washing machine, you can fiddle independently with the drum, the motor, the soap-dispenser and other parts. Unlike the traits of an organism, each of these parts can be detached, manipulated and altered (variants can be generated and tested) without thereby altering or damaging other parts. This independence of parts is important in the construction of any elegant structure that discharges some useful function. Suppose you could not tweak any part of the machine without causing further changes in all the other parts. If you fixed a problem with the soap-dispenser, the knock-on effects on the drum and the motor might leave the machine as a whole no better than before: indeed, sorting out the soap-dispenser is likely to make overall functioning worse. In general, tight integration of this sort will make it very hard indeed to improve a washing machine by tinkering with its parts. Just the same constraint applies to organisms. If they are too tightly integrated, so that any slight change to one trait tends to be accompanied by slight changes to all the others, then it becomes very unlikely that natural selection will be able to produce elegant adaptation. For, even if a chance improvement were to arise in the structure of the eye, it would most likely disrupt the workings of other important organs in the body.

The Harvard biologist Richard Lewontin has summarised this problem by adding that as well as heritable variation in fitness, the further condition of what he terms *quasi-independence* of traits must obtain if complex adaptations are to arise from natural selection (Lewontin 1978). The degree to which a trait is modifiable independently of others – the degree to which it is quasi-independent, or 'modular' – depends on how the development of that trait is controlled as the organism grows from egg to adult. In recent years, biologists have begun to look empirically at the issue of the degree to which organic development is 'modular', and the role that modularity plays in explaining the emergence of complex adaptation. Because those working in this field marry an understanding of the short-term processes of development with a focus on explaining the long-term problems of the evolution of novel traits, their burgeoning field is known as evolutionary developmental biology, or 'evo-devo' for short.

It would be something of a stretch to suggest that Darwin anticipated the major moves of modern evolutionary developmental biology. But Darwin was aware of the need to take into account the relationships between developing traits. Alterations to one trait are often accompanied by modifications to others, sometimes in unexpected ways. In the *Origin* and elsewhere Darwin refers to these relationships as 'correlations of growth'. Darwin understands that the efficacy of selection can be affected by these relationships, a fact which he first points out in the context of animal breeding:

> Breeders believe that long limbs are almost always accompanied by an elongated head. Some instances of correlation are quite whimsical; thus cats with blue eyes are invariably deaf; colour and constitutional peculiarities go together, of which many remarkable cases could be given amongst animals and plants. From the facts collected by Heusinger, it appears that white sheep and pigs are differently affected from coloured individuals by certain vegetable poisons . . . Hence, if man goes on selecting, and thus augmenting, any peculiarity, he will almost certainly unconsciously modify other parts of the structure, owing to the mysterious laws of the correlation of growth.
>
> (*Origin*: 74–75)

Let me summarise. Darwin is aware that one cannot secure the explanation of adaptation by natural selection with an abstract argument showing that only favourable variations are likely to be preserved. Immediately after presenting the abstract précis that I quoted above, Darwin notes that: 'Whether natural selection really has thus acted in nature, in modifying and adapting the various forms of life to their several conditions and stations, must be judged by the general tenour and balance of evidence given in the following chapters' (ibid.: 170). If variation has the wrong character – if it is highly constrained, if it does not permit the quasi-independent modification of single traits – then selection cannot act in the way Darwin says it does. Once we have good reasons to believe that selection is, in general, the agent of adaptation, we can reasonably infer that whenever we encounter an adaptation, variation with the right character must have been available. But Darwin's contemporaries were not in possession of such reasons: he needed to convince them that variation was of the right kind to enable selection to do its work.

Darwin pursued two main strategies to make his case. One was of a direct sort: in 1868 Darwin wrote a vast work entitled *The Variation of Animals and Plants under Domestication*, in which he showed in meticulous detail the range and richness of variation in domesticated species. The second strategy was indirect. Artificial selection – that is, the modification of animal and plant species by human breeders – plays many important roles in the argument of the *Origin*. One of them concerns variation (we will look at others in chapter four). Judicious breeding had given English farmers many improved varieties of cattle, sheep, pigs, cereals and other important species. In this line of work, Darwin tells us: 'The key is man's power of accumulative selection: nature gives successive variations, man adds them up in certain directions useful to him' (*Origin*: 90). The variations nature gives are not themselves under the control of the breeder, he or she merely chooses from what is on offer. By showing that artificial selection has worked in the improvement of breeds, Darwin thereby shows that variation has the characteristics needed to enable selection – whether artificial or natural – to generate adaptation. Artificial selection, too, would be ineffective if variation were not plentiful, or if correlations of growth were so tightly bound as to prohibit the gradual improvement of any one trait. By reminding us of the efficacy of artificial selection, Darwin supports his case for the efficacy of natural selection.

6. SELECTION AND CREATIVITY

Recall that Darwin was much impressed at Cambridge by William Paley's *Natural Theology*. This book exposes the many elegant adaptations of organisms to their environments, in an effort to persuade the reader of the existence of an intelligent God. Paley lingered long over the eye, whose design, he says, eclipses that of the best human engineer. Darwin, like Paley, also puts special weight on accounting for adaptations, referring to especially impressive ones like the eye as 'organs of extreme perfection' (ibid.: 217). But what, in general terms, are adaptations? The most tempting ways to characterise them are in what philosophers call teleological terms, that is, terms which make implicit or explicit reference to goals or ends. Adaptations are traits that are not merely complex in structure – a rubbish dump has a complex structure – they are also well-suited for, or well-directed towards, some purpose. Eyes are good for seeing; hearts are good for pumping blood. One might well wonder whether such teleological description is legitimate, or whether it is instead a hangover from the natural theology that Darwin repudiates. After all, for natural theologians like Paley, organisms are literally complex machines, created by God. It is thus no surprise that natural theologians refer to parts of plants and animals using the language of purpose and design. For Darwin, natural selection replaces the divine artificer. In his *Autobiography* he says that:

> The old argument from design in nature, as given by Paley, which formerly seemed to me so conclusive, fails, now that the law of natural selection has been discovered. We can no longer argue that, for instance, the beautiful hinge of a bivalve shell must have been made by an intelligent being, like the hinge of a door by man. There seems to be no more design in the variability of organic beings and in the action of natural selection, than in the course which the wind blows.
>
> (*Autobiography*: 50)

Why, then, do we find that Darwin continues to characterise the organic world using the language of 'purpose', or of what some trait is 'for', when this vocabulary appears most appropriate for the description of designed artefacts (Lewens 2004)? In the following passage from his 1862 book on orchids, Darwin explicitly discusses plants using the language of machines:

> When this or that part has been spoken of as adapted for some special purpose, it must not be supposed that it was originally

always formed for this sole purpose. The regular course of events seems to be, that a part which originally served for one purpose, becomes adapted by slow changes for widely different purposes . . . On the same principle, if a man were to make a machine for some special purpose, but were to use old wheels, springs, and pulleys, only slightly altered, the whole machine, with all its parts, might be said to be specially contrived for its present purpose. Thus throughout nature almost every part of each living being has probably served, in a slightly modified condition, for diverse purposes, and has acted in the living machinery of many ancient and distinct specific forms.

(Quoted in Browne 2003b: 192–93)

Our problem is to say why Darwin continues to use teleological language for describing adaptations in spite of his rejection of any intentional design overseeing their production. We need to avoid stipulating that teleological descriptions entail the existence of some intelligent agent, or designer. The question is whether, once Darwin banishes intelligent agency from the explanation of adaptation, there remains some legitimate role in biology for the concept of a trait's function or purpose, understood as what the trait is for, or what it is directed towards.

Harvard University's Asa Gray, one of Darwin's regular correspondents, praised him for making teleology respectable in natural history. But Gray thought that the environment selects among variations that are themselves directed, by some agency, towards the welfare of the organisms that bear them. Gray failed to see how else natural selection could play a creative role in the generation of adaptation. Darwin's reply, which appeared at the end of The Variation of Animals and Plants under Domestication, is revealing:

If an architect were to rear a noble and commodious edifice, without the use of cut stone, by selecting from the fragments at the base of a precipice wedge-shaped stones for his arches, elongated stones for his lintels, and flat stones for his roof, we should admire his skill and regard him as the paramount power. Now, the fragments of stones, though indispensable for the architect, bear to the edifice built by him the same relation which the fluctuating variations of each organic being bear to the varied and admirable structures ultimately acquired by its modified descendants . . . Can it reasonably be maintained that the Creator intentionally ordered, if we use the words in any ordinary sense, that certain fragments of rock should assume certain shapes so that the builder might erect his edifice?

(Quoted in Browne 2003b: 293)

Darwin makes it clear that the power of an architect need not depend on any complicit force affecting the supply of raw materials – all that is needed is judicious choice among them. The image of stones falling from a precipice is intended to demonstrate that in Darwin's scheme the raw materials of organic variation are also untailored to their ultimate offices. However, Darwin's reply also draws parallels between selection and conscious design, not as a way of covertly smuggling intelligence into his explanation of adaptation, but as an illustration of natural selection's creativity. If an architect acting on undirected variation can be said to be creative, natural selection deserves to be called creative for the same reasons.

Michael Ghiselin has long argued that there is no room in Darwin's worldview for teleological concepts (e.g. Ghiselin 1994). Why, then, does Darwin so shamelessly name his orchid book (the one I quoted from a few pages ago) On the Various Contrivances by which British and Foreign Orchids are Fertilised by Insects? Ghiselin's answer is that Darwin uses the word 'contrivance' ironically. Ghiselin is certainly right to say that this work is an attempt to undermine the view of orchid species as specially created by a divine designer (Darwin describes the book as 'a flank move on the enemy'). Even so, Darwin's use of terms like 'contrivance' draws attention to natural selection's own creativity. Here we have the beginnings of a justification for teleological descriptions in biology: if designers produce objects with functions or purposes, and if natural selection is similar to a designer in the way it assembles organic traits, then perhaps we can say that organic traits have functions or purposes, too.

We can take this discussion further, moving from a justification of *descriptions* using teleological language ('eyes are for seeing') to a consideration of the related problem of teleological *explanations*. These explanations answer 'Why-questions' – 'Why do we have eyes?' – in teleological terms: 'We have eyes in order to see'. Teleological explanations present us with a general problem, for they seem to explain some state of affairs in terms of a consequence (usually beneficial) it brings. For example, I might explain why Emma is walking to the bakery by saying that she is going there in order to get a croissant. On the face of things, a teleological explanation like this one explains the present in terms of the future. I appear to be explaining what Emma is doing now in terms of something that will only happen after she reaches the bakery. But how can the future explain the past? This would seem to rely on backwards causation, and many philosophers are sceptical of the existence of such a thing. There is an obvious way to solve this problem for Emma, but it won't work so well in the biological context. When I say that Emma is going to the bakery in order to get a croissant, I am really explaining her behaviour in terms of her intention to get a croissant, and this intention precedes her walk.

If teleological explanations always require intentions, then there is little room for teleological explanation in Darwinian biology, save for those cases where we are dealing with animals that have complex mental states. No designer intended that eyes be used for seeing. But perhaps teleological explanations are not, in fact, so demanding. Suppose a selection process is at work on some slow-running wolves. The wolves' environment may be such that *were* these wolves to run faster, they *would* catch more deer. This conditional fact can cause the pack to become composed of faster-running wolves. It is thus legitimate to say that a particular pack of wolves is composed of fast runners *because* running fast helps wolves to catch deer. This explanation is legitimate, but it neither explains the presence of fast runners in terms of a future state of affairs (namely, their catching deer), nor does it explain it in terms of an earlier intention. We might also call this explanation teleological. It is not an explanation of a state of affairs in terms of its consequences, but it is an explanation of a state of affairs in terms of the conditional fact that were that state of affairs to exist, it would have certain consequences. It does not matter that no designer is responsible for wolves' running; we can still say that wolves run fast in order to catch deer. Generalising from this specific example, we can think of natural selection as a process that takes facts about organism/environment relations of the form: 'If members of species S were to have trait T, then they would survive and reproduce better', and converts those facts into the existence of trait T. Thus Darwin licenses us to ask what other organic traits – potentially any trait from the peacock's tail to human emotions – are for, and he shows how we can go about giving answers to those questions (Dennett 1995: chapters eight and nine). Asa Gray was right, but for the wrong reasons.

7. SELECTION AND POPULATION

Let us return again to the case for selection's ability to explain adaptation. Darwin dealt quite well, as we have seen, with the need to establish that variation has the right character. He did not deal so well with what we might call 'populational' worries about selection's efficacy, some of which were put to him forcefully in a review of the Origin that appeared in North British Review in 1867. The review was written by Henry Fleeming Jenkin, Professor of Engineering at the University of Edinburgh ('Fleeming' is pronounced to rhyme with 'lemming', not with 'dreaming').

The historian and philosopher of biology Jean Gayon has studied Jenkin's review closely, and I will borrow heavily from his work in this section (Gayon 1998). Jenkin thought that variation within a species was real, but limited. He believed that there were tight boundaries, which the characteristics of individuals in any given species could never exceed. If this were true, then the wholesale transformation of species in the way Darwin envisaged would be impossible. What is important for our purposes is that Jenkin was prepared to concede, for argument's sake, the unlimited variability of species to Darwin, focusing instead on a quite different set of worries about selection. He raised three problems that are of relevance (my own order of presentation is different from that of Jenkin).

First, Jenkin asks for a fairly precise characterisation of what the relationship of resemblance between offspring and parents (Darwin's 'strong principle of inheritance') is supposed to be. Organisms cannot always resemble their parents perfectly, because often they have two parents who are not perfectly alike. One could answer Jenkin's question in a number of ways. For example, one might say that offspring are roughly half-way between their parents. Or one might hold that offspring always resemble one parent perfectly, and which parent this is is a random matter. Or one might hold that offspring always resemble the better-adapted parent. Jenkin points out that while this last option would enable selection to work efficiently, it cannot be what Darwin has in mind. Why on earth should offspring always end up resembling the better-adapted parent? Some complex mechanism would have to account for this, and Darwin, who says nothing on the matter, would be as guilty of Chambers of appealing to an implausible and mysterious principle in the explanation of adaptation.

Second, Jenkin points out that the efficacy of selection depends on the rate at which variation arises in a population. Darwin tells us that occasional variations arise by chance, and if they are favourable, they will be 'preserved' in the next generation and built upon. But Jenkin shows that chance must be reckoned with at two points. If favourable variations arise very rarely, so that they are scarce in a population, then there is also a good chance that they will not, in fact, be preserved. After all, a huge number of creatures, for one reason or another, will not live to reproduce at all, and these unlucky creatures can include those who are advantaged. Even the swiftest-running wolf in the pack might die before reaching maturity. The more rarely favourable variations arise, the lower the chances are that they will be preserved. Jenkin complains: 'The vague use of an imperfectly understood doctrine of chance has led Darwinian supporters . . . to imagine that a very slight balance in favour of some individual sport [i.e. a rare variation] must lead to its perpetuation' (quoted in ibid.: 94). In fact,

Jenkin argues, one would be mistaken to claim that any favourable variation, however rare, is likely to be preserved by selection; instead, selection can only preserve favourable variations if they arise quite regularly in a population.

Third, Jenkin considers what would happen if those individuals with favourable variations do live to breed. Once again, suppose that favourable variations are rare in natural populations. This suggests that an individual blessed with such advantageous variations will most likely end up mating with another whose endowment is closer to the average. If, for example, swift running arises rarely in a pack of wolves, then a swift runner will probably mate with a wolf of normal running speed. Jenkin then gives Darwin what looks to be a plausible hypothesis of inheritance; namely, that offspring are intermediate in form between their parents. This is now referred to as a hypothesis of 'blending' inheritance. On this assumption, the offspring of our swift-running wolf and its more average mate will run at a speed closer to the pack average than its swift parent. If this little wolf matures and breeds, then once again the chances are that its mate will be an average wolf, and its offspring will be even closer to the average. Over many generations, it seems that the initial rare variation will not be preserved and built upon, but instead will be washed away by repeated cycles of mating and blending.

Darwin took Jenkin's attack very seriously, writing to his friend Joseph Hooker: 'Fleeming Jenkin has given me much trouble, but has been of more real use than any other essay or review' (quoted in ibid.: 85). Jenkin's objections do not show that selection is powerless, and many of the specific implications of his review turn out to be false. But Jenkin raises questions that must be answered if one is to show how favourable variations can be preserved in the way Darwin says they are. What is more, these questions can only be answered by looking at features of populations, features that are best captured using the language of statistics. Darwin's own efforts to confront these problems were poor, primarily because Darwin was far from gifted in mathematics.

Jenkin alerts us, in sum, to three areas of concern. First, the power of selection to produce adaptation depends on how we understand the relationship between parent and offspring. Some readers might be tempted to think that what Darwin needed to make his hypothesis good was a better understanding of the processes that underpin inheritance, perhaps something akin to our own understanding of how the double-helical structure of DNA enables faithful copying of important developmental information. But this is not the best way to describe the situation. What Darwin needed was not primarily a mechanical, causal account of the processes of inheritance, but a set of statistical tools that could express the level of correlation in various characters between parent generations and offspring generations, and which would therefore allow a more rigorous exploration of the patterns of inheritance most conducive to the production of adaptation by selection. Second, Jenkin shows that it is important to take into account the ways in which the frequency at which variations arise in a population can affect the chances of those variations being preserved. Finally, Jenkin shows that likely mate pairings need to be taken into account for they, too, will affect the composition of future generations.

Jenkin, as Darwin's contemporary, offers us a useful insight into a phenomenon that many will find surprising. Although today we perhaps associate Darwin most closely with the hypothesis of natural selection, it was not until the 1920s, 30s and 40s, with the publication of important works by R. A. Fisher, J. B. S. Haldane and Sewall Wright, that natural selection was firmly established in the scientific community as an agent of change and adaptation. Fisher, Haldane and Wright focussed largely on the development of rigorous statistical models of variation, inheritance and selection. Their efforts demonstrate once again that neither the articulation nor the defence of Darwin's explanatory schema is blindingly simple.

8. NATURAL SELECTION THEN AND NOW

After Jenkin, it was no longer enough for selection's defenders to talk in vague terms about the preservation of favourable variations. The problem for advocates of selection becomes a statistical one of analysing how alternative assumptions about the availability of variations, the advantages conferred by those variations, and the level of correlation between parents and offspring, will most likely play out in terms of the composition of populations over several generations. As the primary problem of interest changes to a statistical problem, so a rather different set of emphases are placed upon the understanding of natural selection itself. In this section I want to run over a few of these differences.

First, remember that the existence of a 'struggle for life' is an essential condition for the action of natural selection in Darwin's presentation. He is explicit that this notion of struggle does not require literal battle between members of a species for food or other resources:

> I should premise that I use the term Struggle for Existence in a large and metaphorical sense, including dependence of one being on another, and including (which is more important) not only the life of the individual, but success in leaving progeny. Two canine animals in a time of dearth, may truly be said to struggle with each other, which shall get food and live. But a plant on the edge of a desert is said to struggle for life against the drought, though more properly it should be said to be dependent on the moisture. A plant which annually produces a thousand seeds, of which on average only one comes to maturity, may be truly said to struggle with the plants of the same and other kinds which already clothe the ground.
>
> (*Origin*: 116)

Although Darwin's use of 'struggle' is only metaphorical, it is debatable whether modern evolutionary biologists' conception of selection requires that there be a struggle for existence even in a 'large' sense. The notion of struggle suggests at least that some members of a species will lose if others profit. Plants, for example, might be said to struggle with each other if a plant that takes up nutrients more efficiently than others do thereby deprives those others of scarce nutritional resources. But selection as it is usually understood today does not require scarcity of resources at all. This is best illustrated with an example. Suppose that fast-running wolves enjoy a reproductive advantage over slow-running wolves simply because it takes them less time to catch a deer. This ability to get to food more quickly means they mature at a faster rate, they reach reproductive age earlier, and they have more offspring than the slow runners do because of their earlier start. But if nature is so bounteous that the supply of fast and slow deer is effectively in infinite supply, then slow-running wolves can catch and eat their fill too. Under these circumstances, the numbers of slow-running wolves in a population might expand indefinitely, but if the number of fast-running wolves is expanding even faster, then the frequency of fast-running wolves increases in the wolf population as a whole. This is an instance of selection without struggle. Modern evolution has no essential commitment to the Malthusian view that lies at the heart of Darwin's theory.

Second, in keeping with modern biology's focus on accounting for changes in proportions of traits in populations, selection today is typically characterised as a factor that results in the increase in frequency of one type of trait over another in a population. We can draw on this conception of selection to make clear the positive role that selection plays in the production of adaptation. This is best appreciated with an example. Imagine a local environment that has the capacity to support a pack of 1000 wolves. Let us suppose that there are 100 fast and 900 slow runners in the pack. Finally, let us suppose that a fast wolf is somewhat more likely than a slow wolf to have even better-adapted, extra-fast wolves as offspring. If selection can increase the frequency of fast runners from 10 to 90 per cent, selection thereby greatly increases the chances of an extra-fast wolf being born. By changing the composition of the population, selection plays a positive role in generating adaptive variation. Selection is sometimes misleadingly portrayed as a purely negative process, by which maladaptive variation is weeded out of a population. This negative conception of selection leads some commentators to argue that random mutation is the sole source of adaptive innovation. This picture, in turn, may be responsible for the widespread misconception that there is no significant difference between appealing to selection in the explanation of a complex structure like the eye, and appealing to a random coming-together of matter in explaining the same structure. But natural selection is not a random process in this sense. Natural selection reliably magnifies the representation in a population of partially-adapted forms, and thereby increases the probability of even better-adapted forms being produced in the population by mutation (Lewens 2004: chapter two; Neander 1995).

It is comparatively rare for Darwin to explain adaptation by appealing to the composition of populations in this way. It is not, however, unknown. One of the clearest examples occurs not in Darwin's discussion of natural selection's ability to explain organic adaptation, but in his account of human technical innovation in *The Descent of Man*:

> [if] some one man in a tribe, more sagacious than the others, invented a new snare or weapon . . . the plainest self-interest, without the assistance of much reasoning power, would prompt the other members to imitate him; and all would thus profit . . . If the new invention were an important one, the tribe would increase in number, spread, and supplant other tribes . . . In a tribe thus rendered more numerous there would always be a rather greater chance of the birth of other superior and inventive members.
>
> (*Descent*: 154)

In other words, if an important invention renders a tribe more numerous, the invention thereby increases the chances, merely by increasing the size of the tribe, of further inventive members being born into that tribe who will produce yet more inventions.

Let me close this chapter with a few more remarks on the analogy Darwin draws between natural and artificial selection. Natural selection is vividly portrayed as an agent of greatly superior skill to the common breeder in another famous passage from the *Origin*:

> It may be said that natural selection is daily and hourly scrutinising, throughout the world, every variation, even the slightest; rejecting that which is bad, preserving and adding up all that is

good; silently and insensibly working, whenever and wherever opportunity offers, at the improvement of each organic being in relation to its organic and inorganic conditions of life.

(*Origin*: 133)

This analogy with artificial selection might suggest an image of nature selecting individual organisms to breed or to die according to the beneficence of their variations, just as a sheep breeder picks out, or selects, some individual ewe for further breeding according to the quality of her wool. In contrast with this image, selection today is not understood as a force affecting individual organisms. As we have seen, today's evolutionary theory is concerned with the chances of various types of trait – fast-running, say, or disease-resistance, or camouflage – changing their frequencies in a population. Modern biologists reject the image of selection as acting on individual organisms, in part because they recognise that advantageous trait types need not appear together in individuals. An individual slow-running wolf may have great reproductive success, perhaps because it is also disease-resistant, and well-camouflaged. An individual fast runner may die without offspring, perhaps because it lacks these other advantageous traits. In cases like these, modern biologists do not think of selection as a force that kills our fast runner, or as a force that causes our slow runner to reproduce successfully. Instead, selection is more typically understood as acting on trait types – fast and slow running – according to the average contribution to survival and reproduction of traits of that type, measured across the population as a whole.

Darwin's own descriptions of selection, which are rarely couched in mathematical language, make it hard to say to what degree his conception differs from the modern one. On the one hand, Darwin speaks of natural selection aiming at the improvement of *each organic being*, and of *a plant* struggling against drought. On the other hand, he also tells us that selection scrutinises *variations*, not individual organisms. However we should situate Darwin himself, modern biology rejects the image of natural selection as a scrutineer of individual organisms, in spite of what the analogy with the long arm of the breeder might suggest.

SUMMARY

Darwin was not the first to claim that distinct species evolved from common ancestors. He was not the first, that is, to defend the hypothesis of evolution. But he felt that a defence of the evolutionary hypothesis would not be satisfactory unless he could also explain how species came to be so well adapted to their environments. This was the problem that the hypothesis of natural selection was invoked to solve. Consequently, Darwin did not merely offer an abstract argument linking inheritance and differences in reproductive successes among species members to the accumulation of favourable variation in those species. He understood that he if he were to defend the claim that natural selection explained adaptation, he would need to give an account of the general character of variation. He was also aware of a second set of problems for selection, raised by worries about how likely it was that rare advantageous variations would be 'preserved' in populations. Although Darwin dealt well with the first set of worries, his mathematical naivety meant that he was not in a position to deal so effectively with the second set. These worries became the focal point of later work in mathematical population genetics, the theoretical core of today's evolutionary biology. The problems faced by natural selection explanations show that the principle of natural selection is not as simple as some commentators suggest. It takes considerable empirical and theoretical work to show that natural selection is capable of explaining adaptation. Darwin did not complete that job, although he did start it.

FURTHER READING

For Darwin's own thoughts on natural selection, the first seven chapters of the *Origin* are probably the best places to look, particularly chapters one, three and four.

Secondary literature on natural selection is voluminous. A useful historical overview of evolutionary ideas is contained in:

Bowler, P. (1984) *Evolution: The History of an Idea*, Los Angeles: University of California Press.

Two philosophical approaches to natural selection which are also historically sensitive are:

Gayon, J. (1998) *Darwinism's Struggle for Survival*, Cambridge: Cambridge University Press.

Depew, D. and Weber, B. (1996) *Darwinism Evolving: Systems Dynamics and the Genealogy of Natural Selection*, Cambridge, MA: MIT Press.

Two works that focus on natural selection primarily as it is understood in modern biology, but which cover far more than natural selection alone, are:

Sober, E. (1984) *The Nature of Selection: Evolutionary Theory in Philosophical Focus*, Cambridge, MA: MIT Press.

Dennett, D. C. (1995) *Darwin's Dangerous Idea: Evolution and the Meanings of Life*, London: Allen Lane.

On the continuity between natural theology and modern biology, see:

Lewens, T. (2005) 'The Problems of Biological Design', in A. O'Hear (ed.) *Philosophy, Biology and Life*, Cambridge: Cambridge University Press.

Finally, a useful collection of articles on teleology in modern biology is:

Allen, C., Bekoff, M. and Lauder, G. (1998) *Nature's Purposes: Analyses of Function and Design in Biology*, Cambridge, MA: MIT Press.

Three

Species

1. HUMAN NATURE, SQUID NATURE, APPLE NATURE

For many readers, a particular set of views about human nature will probably spring to mind when they think of Darwin. This package is likely to include the claims that human nature is something that was shaped by natural selection many thousands of years ago, something that exists independently of prevailing human social environments, something all humans share, probably something that is innate, and something that is largely immune to manipulation by social reform.

Does Darwin really condone such an image of human nature? Subsequent chapters of this book look at this package of views in detail. But some modern-day Darwinians argue that species, whether they are apples, squid or humans, are not the right sorts of things to have natures at all (e.g. Hull 1998). These Darwinians tend to use four arguments to support their sceptical case. In their view, Darwin teaches that a species is a branch (or a twig) on the Tree of Life. Just as a real branch of a real tree can be composed of diverse parts with little in common, so there is no reason to suppose that all the individual organisms that are parts of a species have commonly-held properties. Hence there is no reason to think that there is such a thing as human nature, understood as a set of characteristics which define what it takes to be a member of the human species. Second, Darwin teaches us that species are mutable. The essence of the evolutionary view is that species change over time, with the result that the individuals who are members of some species at one time can be quite different from individuals who are members of the same species at a later time. Here, again, there is no room for a conception of human nature as some set of characteristics all our species' members share. Third, Darwin teaches us that variation is the fuel of evolution. Rare anomalies in a species, if they are advantageous, can over time become common, while characteristics that are common may disappear. So there is no sense in picking out some set of characteristics common at a moment in time as emblematic of 'human nature', for these can disappear before the human species does, and they can be replaced by characteristics that do not yet exist. Fourth, Darwin teaches us that diversity within a species need not be ephemeral. Natural selection does not always work to make just one of the available variant forms of a trait universal in a species. Selection can fashion populations which feature a stable mixture of alternative variants. So once again, we should not expect to find a single form that exemplifies the 'nature' of a species.

Such are the consequences said to follow from a properly Darwinian view of species. So, what is a species?

2. THE TREE OF LIFE

Let us begin this enquiry by examining Darwin's views about how species are formed. At the end of the Origin's chapter on 'Natural Selection', Darwin paints an arboreal picture of the structure of organic life:

> The affinities of all the beings of the same class have sometimes been represented by a great tree. I believe this simile largely speaks the truth. The green and budding twigs may represent existing species; and those produced during each former year may represent the long succession of extinct species. At each period of growth all the growing twigs have tried to branch out on all sides, and to overtop and kill the surrounding twigs and branches, in the same manner as species and groups of species have tried to overmaster other species in the great battle for life . . . Of the many twigs which flourished when the tree was a mere bush, only two or three, now grown into great branches, yet

survive and bear all the other branches; so with the species which lived during long-past genealogical periods, very few now have living and modified descendants. From the first growth of a tree, many a limb and branch has decayed and dropped off; and these lost branches of various sizes may represent those whole orders, families, and genera which have now no living representatives, and which are known to us only from having been found in a fossil state . . . As buds give rise by growth to fresh buds, and these, if vigorous, branch out and overtop on all sides many a feebler branch, so by generation I believe it has been with the great Tree of Life, which fills with its dead and broken branches the crust of the earth, and covers the surface with its ever branching and beautiful ramifications.

(*Origin*: 171–72)

Note that Darwin retains a certain modesty here. With the luxury of data from molecular biology that Darwin did not, of course, have access to, most of today's biologists would reckon that all of plant and animal life traces back to one single point of origin. Darwin, by contrast, withholds judgement on the issue of exactly how many original life forms (hence how many trees) we should recognise: ' . . . I cannot doubt that the theory of descent with modification embraces all the members of the same class. I believe that animals have descended from at most only four or five progenitors, and plants from an equal or lesser number' (ibid.: 454). Examples of *classes* include the mammals (Mammalia) and the insects (Insecta). Since Darwin says he thinks all animals can be traced back to four or five ancestors at most, and since there are far more than four or five classes of animal, we must assume that he believes all members of each *phylum* form a tree. A phylum is a more comprehensive grouping of species, such as Chordata – a group which includes the vertebrates, and which is typically defined by reference to the possession of a cartilaginous rod that runs down the back – or Arthropoda – the phylum comprising, among other things, insects and crustaceans.

The Tree of Life metaphor illustrates Darwin's view of natural history as a process of 'descent with modification': the various species of a class all have some distant common ancestor, of which they are the altered descendants. Logically speaking, one could accept that life has a tree-like structure while denying any important role in life's history to the process of natural selection (Waters 2003). Darwin's view, however, is that natural selection is both the primary agent of adaptation and the primary agent by which new species are formed.

Darwin thinks that new species are formed by one of two mechanisms. The easiest to appreciate relies on *geographical isolation*. A population of individuals of one species splits into two groups, which are geographically cut off from each other. This could happen for numerous reasons; a small number of migrating birds might be blown off course and separated from the main flock to settle in a new territory, or rising sea levels might lead to the creation of an island populated by stranded mammals. When a single population splits into two like this, the environmental demands on the two groups are likely to differ. Natural selection then results in the generation of differences in the makeup of the groups. This may not be enough by itself for the creation of new species, because if the geographical isolation in question is short lived (the birds remain isolated for a couple of generations before meeting a large flock from their original species the flood waters recede permitting the mammals to rejoin the mainland), then breeding might start up again between the two groups, and the further accumulation of differences will be stalled. But if the differences accumulated during a period of isolation are enough to make it impossible for the members of one group to mate successfully with the other, or even if they make it very unlikely that a member of one group would attempt to mate with a member of the other (perhaps because selection has affected some indicator of mate suitability), then the two groups can remain effectively sealed off one from the other even if they return to live in the same geographical area. Once this happens selection can cause the two groups to diverge even more until they form unquestionably distinct species.

Modern biologists continue to view this mechanism of species formation as important. Many are more sceptical of Darwin's second mechanism, which relies on his *principle of divergence of character*. Like his appropriation of Malthus, this is an extension of economic reasoning into the biological realm. Adam Smith had claimed that competition will be most intense between individuals serving identical markets. Darwin concludes that in the economy of nature, no less than in human affairs, advantage will come to those who open new markets:

> . . . the more diversified the descendants from any one species become in structure, constitution, and habits, by so much will they be better enabled to seize on many and widely diversified places in the polity of nature, and so be enabled to increase in numbers.
>
> (*Origin*: 156)

A single physical environment offers many ways to organisms of making their living – in today's language, one environment contains many niches. Individual members of a species with unusual anatomy or even unusual preferences (an insect whose proboscis allows it to access the nectar of peculiarly shaped flowers, a bird which chooses to feed on strange seeds) can thrive owing to a lack of competition in their niche. In this way, says Darwin, a single species containing largely uniform members will diversify so as to feature an array of individuals of different types, adapted to different niches. At this stage, individuals of different types may be physiologically capable of breeding together, but for one reason or another they do not do so. (Perhaps their different feeding habits mean they rarely encounter each other, perhaps their different anatomies mean they are not attractive to each other.) Selection then cements these differences further so that interbreeding is impossible, and distinct species are formed.

Why think that we can move from a single species that contains diverse individuals to a multiplicity of distinct species? When the process of diversification begins, all the members of the species can mate with each other, and they live in a single physical environment. We might expect, then, that individuals partially adapted to one niche will often mate with individuals partially adapted to other niches, with the result that the process of divergence will stall. It is for reasons like this that modern biologists remain divided on the efficacy of the principle of divergence for speciation (Coyne and Orr 2004).

Darwin's two mechanisms – geographical isolation and the divergence of character – have a good deal in common. In both cases, species are formed when natural selection accentuates differences between two or more commonly found forms or *varieties* within a single species, with the result that varieties which at first do not mate with each other (because, for example, they are geographically isolated from each other, or because their different habits keep them apart) later cannot mate with each other. Hence Darwin's assertion that: 'according to my view, varieties are species in the process of formation, or are, as I have called them, incipient species' (*Origin*: 155).

3. BUTCHERING NATURE

There is a long-standing philosophical view that tells us that the job of the sciences is to 'carve nature at the joints'. This view presupposes, of course, that nature has joints for our cleavers to find. Looking across the physical sciences, one might think this view is often justified. The periodic table, for example, looks like a good representation of the different kinds of chemical stuff that exist independently of our inquiry: our recognition and tabulation of the distinct elements constitutes a successful piece of scientific butchery.

Not all sciences are the same. One is likely to think that this chemical taxonomy – the classification of elements – is quite different from some of the practices of astronomical taxonomy, especially the classification of stars into constellations. We group stars together in the night sky in order to enable navigators to recognise and remember its features for the purposes of orientation and communication. Stars in the same constellation need not be anywhere near each other in space, only in space as it appears to the Earth-bound observer. So only in the most strained sense does the practice of carving up the night sky mirror important divisions in the universe itself.

What of species? Does the biological world contain joints at which we can carve? Or do we merely project distinct species onto the multiplicity of individual organisms for our own sake, to facilitate cataloguing, communication and other human practices? Are species more like elements, or constellations? Let us use the term *realism* for the view that species are, in some sense, natural parcels of the biological world that exist independently of human investigation. Let us use the term *nominalism* for the view that species are mere artefacts of our convenience, which reflect nothing more than our own efforts to put nature's diversity into tractable order.

I have introduced the terms *realism* and *nominalism* because they are in such common currency that it is hard to get by without them. But unless handled with care they can make debates confused and simplistic. Rather than beginning this discussion with an abstract attempt to clarify these terms, let us look directly at Darwin's work on species, and use this to sharpen the various senses in which we might be realists or nominalists.

In a few places Darwin seems to commit himself to a kind of nominalism about species. The most convincing statement of this kind comes in chapter two of the *Origin*:

> . . . I look at the term 'species', as one arbitrarily given for the sake of convenience to a set of individuals closely resembling each other, and that it does not essentially differ from the term variety, which is given to less distinct and more fluctuating forms. The term variety, again, in comparison with mere individual differences, is also applied arbitrarily, and for mere convenience sake.
>
> (ibid.: 108)

Darwin is telling us that some individual organisms, although differing one from the other, are generally quite similar. Dogs resemble each other in various respects, for example. A group of similar organisms can be subdivided once again so as to pick out sub-groups that resemble each other in more specific ways. The terriers might constitute such a sub-grouping within the dogs. Groupings of the first kind are species; groupings of the second kind are varieties. Darwin also believes that when we look at groups of resembling organisms, it will sometimes be quite clear as to whether we should call the group a species or a variety. He thinks of species as 'strongly marked and permanent varieties': the notion of permanence is important here, for it explains why Darwin makes use of the ability to reproduce together as an indicator of when two organisms are members of the same species. If organisms of two different forms cannot reproduce when placed together, then this suggests that the differences between these forms are likely to remain permanent, for mating will not lead them to disappear. But, since species are formed when varieties are altered by degrees to become 'strongly marked and permanent', we should not expect to find any clear moment when a variety has become distinct enough, or the difference between it and another variety is likely to last long enough, that it should be termed a species.

The nominalism that Darwin defends here is of a weak kind. Consonant with realism, he tells us that both species and varieties are groups of organisms that have properties in common. That they have properties in common is, in Darwin's view, true independently of human interests. Equally consonant with realism, he tells us that we can often decide with certainty whether some group of resembling organisms is a species or a variety. Darwin's nominalism extends only as far as a scepticism about a very sharp boundary between varieties and species. This is, of course, a consequence of his view that species are formed when varieties become 'strongly marked and permanent': we may need to decide arbitrarily whether the status of species has been attained by a variety. But note how mild this nominalism is. We believe these days that radioactive decay leads to elements of one kind being transformed into elements of another. Can we say at what moment a decaying atom of Uranium-238 becomes an atom of Thorium-234? Scepticism regarding a precise boundary between the two hardly amounts to the view that the periodic table represents no natural ordering of the elements (Sober 1980). Darwin's scepticism about the variety/species distinction is similar.

If Darwin is right about how species are formed, we should expect there to be irresolvable bickering among naturalists regarding whether some groups of resembling organisms should be termed species, or mere varieties. And this, Darwin says, is just what we do find:

> On the view that species are only strongly marked and permanent varieties, and that each species first existed as a variety, we can see why it is that no line of demarcation can be drawn between species, commonly supposed to have been produced by special acts of creation, and varieties which are acknowledged to have been produced by secondary laws.
>
> (*Origin*: 443)

But this does not undermine a more general realism about species, and Darwin often expresses his views about taxonomy in a way that explicitly distances him from nominalism:

> From the first dawn of life, all organic beings are found to resemble each other in descending degrees, so that they can be classed in groups under groups. This classification is evidently not arbitrary like the grouping of the stars in constellations.
>
> (Ibid.: 397)

Here Darwin is not merely talking about species, but about so-called 'higher taxa' (e.g. families, classes, phyla) too. In this paragraph, he is asserting that the system of classification of groups under groups – a hierarchical theory of classification – does indeed sort organisms by closeness of resemblance. This suggests that Darwin is a realist regarding the existence of what we might call the 'clustering' of organisms into groups of resembling individuals. Much later, a view of this kind would be clearly expressed by the eminent evolutionary geneticist Theodosius Dobzhansky:

> Although individuals, limited in existence to only a short interval of time, are the prime reality with which a biologist is confronted, a

more intimate acquaintance with the living world discloses a fact almost as striking as the diversity itself . . . A multitude of separate, discrete, distributions are found . . . Each array is a cluster of individuals which possess some common characteristics. Small clusters are grouped together into larger secondary ones, these into still larger ones, and so on in an hierarchical order.

(Dobzhansky 1951: 4)

Dobzhansky, too, articulates a view of this classification that acknowledges the manifest fact that the system of classification we use is man-made, while asserting, as Darwin seems to believe, that it reflects a true hierarchy in nature:

For the sake of convenience the discrete clusters are designated races, species, genera, families, and so forth . . . Biological classification is simultaneously a man-made system of pigeonholes, devised for the pragmatic purpose of recording observations in a convenient manner, and an acknowledgment of the fact of organic diversity.

(Ibid.: 5)

An additional dose of realism comes from Darwin's ambition to produce what he calls a 'Natural System'. He is not content merely to class organisms according to their resemblances; he also attempts to give an explanation for why organisms should fall into a hierarchical pattern of classification. Darwin describes himself as a 'philosophical naturalist', and this means a naturalist whose classification should not merely, to borrow Dobzhansky's words, fit 'the pragmatic purpose of recording observations in a convenient manner', but one who looks to give some rationale for nature's mode of organisation. More specifically, this rationale should be based on natural laws (Rehbock 1983: 4). Hence Darwin remarks, importantly, that although the hierarchical classification of group under group properly organises creatures into those that resemble each other to greater and lesser degrees:

I believe something more is included; and that propinquity of descent,—the only known cause of the similarity of organic

beings,—is the bond, hidden as it is by various degrees of modification, which is partially revealed to us by our classifications.

(*Origin*: 399)

It is because species are descended from each other that the individuals within a species resemble each other closely, individuals
within a genus less closely, and individuals within a family less
closely again. A 'Natural System' of classification thus places
organic beings into a hierarchical system of resemblance, while at
the same time revealing the ground of that resemblance in terms
of a set of genealogical relationships. Darwin's units of classification are real units because, not merely are they units of genuinely
similar organisms, there is also a reason – common ancestry – for
their similarity. Darwin understands that we could use any resemblances we like to sort organisms, but a taxonomic system that
reflects genealogical relations is the only one that simultaneously
explains and makes manifest large classes of resemblances. This
view is most clearly expressed in the *Descent of Man*, and expresses a
strong taxonomic realism:

> Classifications may, of course, be based on any character what-
> ever, as on size, colour, or the element inhabited; but naturalists
> have long felt a profound conviction that there is a natural
> system. This system, it is now generally admitted, must be, as far
> as possible, genealogical in arrangement . . .

(*Descent*: 174)

4. INDIVIDUALS AND KINDS

We have not yet addressed a question that holds considerable
interest for some philosophers these days. This is the question of
whether species are *individuals* or *kinds*. Although these terms are
used in everyday language, here I am using them in a technical
sense. A hasty sketch of their technical meanings will have to
suffice. Some fairly uncontroversial examples of kinds might
include *things with a mass of three kilograms*, and *lumps of pure gold*. A kind
is a set of objects, united by shared properties. Members of kinds

can therefore be far flung; there are doubtless objects scattered all over the universe with a mass of three kilograms. An individual, on the other hand, is an object or event, with a beginning and end in time, and with physical boundaries. Some fairly uncontroversial examples of individuals might include Lord's Cricket Ground, the lamp on my desk, Tony Blair and The 2006 World Cup Final. There might not be very precise moments at which these things come into existence, but we can always give reasonably good approximations. The same goes for their physical boundaries. (Where, exactly, does Tony Blair finish and the stuff around Tony Blair begin? Somewhere around his skin.)

Although the question of whether species are kinds or individuals is extremely abstract, it nonetheless has implications for the reality of human nature. Non-biological readers may think it obvious that species are kinds. 'What is a species,' one might ask, 'other than a set of organisms with properties in common?' Such a view is likely to lead to the conclusion that all species have natures; human nature is simply the collection of properties, which together determine what it takes for an organism to be classified as a member of the species Homo sapiens. Someone who thinks species are individuals is likely to be more sceptical about human nature. We can appreciate this scepticism by considering a non-biological individual, such as the lamp sitting on my desk. It has several parts – the base, the bulb, the shade, the switch. These parts have very little in common with each other. They are all united in the lamp, but not because they share some 'lamp nature'. They are parts of the same object in virtue of the relations they stand in to each other, not in virtue of their shared properties. If species are also individuals – if, that is, they are entities with a beginning and end in time, which have organisms as parts – then there is no more need for their parts to have properties in common than there is for the parts of a lamp to have properties in common.

We must be careful not to be too hasty in claiming consequences for the species-as-individuals view. Although the parts of my lamp have little in common with each other, other individuals – the serving of rice I had for dinner last night, for example – contain parts that resemble each other very closely. So even if we were to conclude that species are individuals, we might still conclude that some, or most of them, have uniform parts. *Homo sapiens* might turn out to be an individual whose parts – particular human beings – are largely alike. So although the species-as-individuals view is compatible with the denial of species natures, it does not force such a denial upon us.

In order to make progress with the question of whether species are individuals or kinds, we need to examine philosophical conceptions of kinds in a little more detail. I have not yet addressed the question of what makes kinds *natural*. *Natural kinds*, on many philosophical views, are the basic sorts of things that science seeks to find and characterise. Once again, the chemical elements provide canonical examples. Although the set of things with a mass of three kilograms is indeed a collection of objects with a shared property, many will not regard it as particularly 'natural', comprising as it does various animals and plants, as well as bowls of pasta served in the USA, several rocks and some large books. This is a motley assortment of objects indeed, which one might well contrast with more 'natural' kinds, such as the set of lumps of pure gold. The latter set comprises objects which have many important properties in common, relating to such diverse features as electrical conductivity, density and melting point. The fact that many properties of scientific interest come together in lumps of pure gold makes these objects important ones for science to classify and investigate – more so than objects with a mass of three kilograms. For this reason, while one might regard *having- mass-three-kilograms* as a *property* of scientific importance (in mechanics, for example), one might resist calling the set of objects that possess this property a *natural kind*.

The philosopher Richard Boyd has defended a view of natural kinds that draws on these types of observations (e.g. Boyd 1991). Boyd thinks we should recognise gold as an example of a natural kind because, as we have already seen, several properties of scientific interest cluster together in lumps of pure gold. What is more, this is no accident. It is because of facts about, for example, the microstructure of pure gold, that lumps of gold have all of these different properties. Natural kinds, in Boyd's view, are collections of objects in which the same group of properties reliably cluster together, and for which some factor explains this co-occurrence. He refers to whatever plays this explanatory role as a 'homeostatic mechanism'. Hence, says Boyd, natural kinds should be understood as *homeostatic property clusters*.

Armed with these various pieces of technical terminology and theory, we can now look at the issue of whether species are kinds or individuals. The species-as-individuals view was originally advocated by the biologist Michael Ghiselin (1974) and the philosopher David Hull (1978). Their view is that a proper Darwinian view of species tells us that a species is a twig or a branch on the Tree of Life. Our own species – *Homo sapiens* – is, like Tony Blair, something with a beginning and (eventually) an end in time, as well as physical boundaries. Those boundaries have changed over time as the species has expanded its range over the planet, just as Tony Blair's physical boundaries change over time as his waistline expands and contracts. And the species has changed in other ways as it has evolved under selection, just as Tony Blair has changed over time with the ravages of office. These considerations, and others, make it most appropriate to think of a species as an individual.

This all stands in stark contrast to a traditional view among philosophers of science who regularly mention species – for some reason, tigers are a favourite example – alongside gold and water in their standard examples of natural kinds. But, say the likes of Hull and Ghiselin, species should not be understood as sets of similar objects at all. Two distinct tigers are parts of the same species not because they resemble each other, but because of the relations they stand in to each other. Specifically, they need to be related to each other genealogically. If species were kinds, then if an object sprang into existence somewhere on a distant planet that looked and behaved (on the insides and outsides) just like a tiger, then it would be a tiger. However, say Ghiselin and Hull, whatever this alien beast is, it is not a tiger unless it is genealogically related to our terrestrial tigers.

Does Darwin think that species are individuals, or kinds? In some ways this is a silly question, because the terms 'individual' and 'kind' are modern inventions, which feature in debates in which Darwin is not a participant. Even so, the passages from the *Origin* that I have cited in this chapter up to now seem explicitly to invoke a conception akin to that of species as kinds. Thus, to repeat a passage we have already seen: ' . . . I look at the term "species", as one arbitrarily given for the sake of convenience to a set of individuals closely resembling each other . . . ' (*Origin*: 108). The resemblances in question need not be restricted to external features that are perceptually obvious. They might also include commonalities of internal structure, or developmental processes. In the last section we saw that Darwin claims that shared ancestry is what explains these close resemblances between species members. In this respect, Darwin's view seems strikingly close to a version of the homeostatic property cluster theory of natural kinds. Species are natural kinds, for they are sets of organisms resembling each other in numerous respects, and the fact of these resemblances being found together is explained by shared ancestry. This is precisely the view of species that has recently been defended by the philosopher of biology Paul Griffiths (1999).

Alas, it is not so easy to conclude that Darwin espouses the view that species are kinds. Other passages suggest that he thinks species are more like individuals. He argues strongly both in the *Origin* and *Descent* that: 'all true classification is genealogical' (*Origin*: 404). A good classification, on this view, is simply an accurately drawn Tree of Life. This suggests that Darwin thinks species are individuals, because portions of the tree of life are individuals in the sense that we have outlined. They are objects that come into existence at a time and disappear when they become extinct. They have organisms for parts, and these organisms are all parts of a single object in virtue of their being bound together by relations of descent.

It is hard to attribute Darwin's view decisively to the kinds-camp or the individuals-camp, not merely for the obvious reason that 'kind' and 'individual' are modern terms of art, but also for the more substantial reason that Darwin believes common ancestry explains and produces resemblances between organisms. The best way to uncover and illuminate shared resemblances between individual organisms requires an understanding of their genealogical relations. So if species are understood as kinds, we can achieve the most informative taxonomy by looking to the historical relationships between organisms. Conversely, if species are understood as chunks of the Tree of Life, then, even though they are individuals, their parts will, as a matter of fact, consist in sets of closely resembling organisms. Darwin illustrates his belief that resemblance and genealogy ('affinity' and 'filiation' in his terms) go hand-in-hand by defending an analogous view of the classification of languages:

> . . . the proper or even the only possible arrangement would be genealogical; and this would be strictly natural, as it would connect together all languages, extinct and modern, by the closest affinities, and would give the filiation and origin of each tongue.
>
> (*Origin*: 406)

The only way we could bring Darwin out into the open would be by pressing him on what he would say in the unlikely circumstances that, for example, a bear gave birth to what looks like a kangaroo. The species-as-kinds view would say that the creature is indeed a kangaroo, albeit an extremely improbable kangaroo. This is because it resembles kangaroos. The species-as-individuals view would say that the creature is not a kangaroo, but a bear, albeit an extremely odd-looking bear. This is because it is part of the bear portion of the tree of life, in virtue of being the offspring of a bear. Darwin considers precisely this example, and his answer tells us whether he is with the kinds-camp or the individuals-camp: ' . . . what should be done if a perfect kangaroo came out of the womb of a bear? . . . The whole case is preposterous, for where there has been close descent in common, there will certainly be close resemblance or affinity' (ibid: 408). The question is too silly to merit an answer, and as a result Darwin refuses assimilation to either school.

Independent of where Darwin stands on the question, what are the relative merits of the species-as-individuals view compared with the view of species as kinds? The most plausible view of species as kinds is that which I attributed to Griffiths, and which I argued is partially prefigured in Darwin's own work. This is the view according to which species are indeed sets of resembling organisms, where it is common ancestry that explains these resemblances. One might think that this view cannot stand up to scrutiny, simply on the grounds that species are so variable. Take any property you like; it seems unlikely that it will pick out all and only members of a single species. Not all humans are bipeds: some have only one leg. Not all humans are rational: some are not capable of thought at all. This variability of species in every respect is something that Darwin himself repeatedly brings to our attention: 'I am convinced that the most experienced naturalist would be surprised at the number of cases of variability, even in important parts of structure . . . ' (ibid.: 102). This means that any supporter of the species-as-kinds view will have to argue for a mild version of that position. They obviously cannot maintain that species are sets of identical organisms, nor even that they are sets of organisms which all have a number of properties in common, but only that they are sets of roughly similar organisms, all things considered. This is compatible with the view that every member of every species is unique.

Even this rather watered-down view of species as kinds faces problems. Many – perhaps most – species are highly *polymorphic*; that is, they come in a variety of forms (Ereshefsky and Matthen 2005). (Darwin's view that species derive from what were once distinct *varieties* relies on the existence of polymorphic species.) Some level of polymorphism should be expected as a result of the variation which Darwin believes is constantly arising in every species. More interesting for our purposes are the ways in which selection can promote and maintain the existence of distinct polymorphic forms. Most obviously there are sex-differences. In birds of paradise, for example, the males are highly coloured, while the females are comparatively drab. There are other ways in which selection can maintain polymorphisms. Consider the question of whether it is better for an animal to be aggressive or passive when it competes with others for mates. The answer depends, in part, on the nature of other members of the population. When only a few are aggressive and most are passive, aggression can pay handsomely, for the aggressor initiates plenty of conflicts which are won by a walkover. When only a few are passive and most are aggressive, aggression can be a liability, for the aggressor continually gets into drawn-out, energy-sapping fights with other aggressive individuals. Hence, as selection makes aggression more common in a population, the fitness of passivity increases until eventually it is the fitter strategy. But as selection makes passivity more common, the fitness of aggression comes to exceed that of passivity. The end result is a stable mixture of the two behavioural strategies. This is a rough sketch of the classic 'Hawk–Dove' model, first developed by biologists John Maynard Smith and Geoffrey Parker (1976).

Abstract thinking of this general form suggests that selection will sometimes cause a mixture of behavioural strategies to co-exist in a species. It remains neutral, however, on the question of how these mixtures are realised. One possibility is that some organisms adopt one strategy, while others adopt another. A second possibility is that all organisms adopt identical 'mixed strategies', which combine (in the case we just looked at) aggression and passivity. If 'mixed strategies' were the general rule, then one might continue to claim that selection tends to produce uniform populations. But there is plenty of research to indicate that in the natural world selection often produces variegated populations whose members use different strategies. Reproductive behaviour is a domain in which there are several well-documented examples of this sort. In the marine crustacean species *Paracerceis sculpta*, for example, males use one of three strategies (Shuster and Wade 1991). Large ones defend 'harems' of females, sequestering them inside sponges. Smaller ones use a strategy of mimicry. They look like females, and they also copy female behaviour, causing the larger males to allow them into the harem. Finally, tiny ones slip into harems unnoticed. Over time, the average fitnesses of these strategies are identical, and selection maintains all three. (I have borrowed this lovely example from Buller 2005: 43.)

The existence of widespread polymorphism need not threaten the view of species as kinds. Species can still be understood as sets of roughly similar organisms all things considered, even if species members show significant differences in particular traits (perhaps relating to plumage or reproductive behaviour). Suppose it turns out, though, that sexual selection causes the males of several closely-related bird species to acquire very different suites of gaudy plumage and showy behaviours, where each suite of plumage and behaviour is particular to the males of a single species. And suppose the females in these closely-related species, because they are not subject to strong sexual selection, do not diverge from each other all that much. The result is that the females of any one species resemble the females of the other related species far more closely than they resemble the males of their own species. Under circumstances like these, it is hard to regard any one species – a collection of males and females combined – as a natural kind. That does not mean that there are no kinds to be found here. It would appear that all the female birds from all the related species form a kind. All the males and females from all the related species might form another kind. But it is hard to see what form of all-things-considered resemblance could draw together a set of organisms comprising only males and females of single species. I conclude that the species-as-kinds view is hostage to the form that poly-morphism happens to take in the natural world.

5. POPULATION THINKING AND TYPOLOGICAL THINKING

Ernst Mayr, one of the most important biologists of the twentieth century, and also an influential contributor to the history and philosophy of biology, consistently maintained over his long career that one of Darwin's greatest contributions to thought was of a philosophical character. He argues that Darwin replaced one way of thinking about nature, which Mayr calls 'typological thinking', with another, which Mayr labels 'population thinking' (Mayr 1976). In this section we will say a little about what this means in the context of species.

In a short passage from a now classic article, Mayr introduces the main contours of the population/typological distinction, specifically defining typological thinking with respect to Plato's doctrine of 'forms' ('*eidos*' in the Greek). Plato thought that in addition to individual virtuous acts, or individual beautiful objects, there was some further, abstract, eternal 'form' of virtue, or beauty, manifested in these worldly acts and objects to a greater or lesser extent. Mayr takes the typologist in biology to hold an analogous view, positing an ideal specimen or 'form' for each biological species:

> According to [typological thinking], there are a limited number of fixed, unchangeable 'ideas' underlying the observed variability, with the *eidos* (idea) being the only thing that is fixed and real, while the observed variability has no more reality than the shadows of an object on a cave wall, as it is stated in Plato's allegory. The discontinuities between these natural 'ideas' (types), it was believed, account for the frequency of gaps in nature . . . Since there is no gradation between types, gradual evolution is basically a logical impossibility for the typologist. Evolution, if it occurs at all, has to proceed in steps or jumps.
>
> (Ibid. 1976: 27)

This is typological thinking. Darwin, on the other hand, is a population thinker:

> The assumptions of population thinking are diametrically opposed to those of the typologist. The populationist stresses the uniqueness of everything in the organic world . . . All organisms and organic phenomena are composed of unique features and can be described collectively only in statistical terms. Individuals, or any kind of organic entities, form populations of which we can determine only the arithmetic mean and the statistics of variation. Averages are mere abstractions; only the individuals of which populations are composed have reality.
>
> (Ibid.)

Finally, Mayr draws a direct contrast between the two positions:

> The ultimate conclusions of the population thinker and the typologist are precisely the opposite. For the typologist, the type (*eidos*)

is real and the variation an illusion, while for the populationist the
type (average) is an abstraction and only the variation is real. No
two ways of looking at nature could be more different.

(Ibid.)

As I see it, the best way to understand the typological/population
distinction is as a disagreement over how to explain the patterns
of similarity and difference between individual organisms. This
means explaining what we have already called the 'clustering' of
organic forms. Organic forms cluster together in what we might
term 'morphospace' – the abstract space of possible organic
forms. We place individual organisms into this space according to
how closely they resemble each other. Highly similar organisms
are very close together; highly dissimilar organisms are very
distant. Some areas of this space are empty: there are no six-
legged elephants. And others are quite densely populated: very
many organisms have a dog-like appearance. In brief, the occu-
pants of morphospace are clumped together in patches, and this
clumpiness is something that needs explanation.

The typologist thinks the tight clustering in morphospace of
individual dogs (even if they are all strictly unique), and the
emptiness in morphospace of the six-legged elephant area, or the
unicorn area, is a manifestation of the fact that some 'doggy' type
underlies the dog area, while no type underlies the unicorn or six-
legged elephant area. Mayr consequently misleads by stressing the
population thinker's belief in the uniqueness of every individual
organism: the typologist certainly is not committed to the absurd
view that all dogs are identical, and he could even agree with the
population thinker that no two organic beings are alike (Sober
1980).

What are these types supposed to be? We do not need to write
them off as unaccountably mystical. Moving away from biology
for a moment, something like a typological explanation seems
appropriate when we try to understand why some crystal structures are seen frequently, while others are not seen at all. We can
take reference to types here to be shorthand for sets of physical
facts that make some crystalline forms stable, others unstable.

Perhaps we can think of organic types in a similar way. The typologist claims that only a few basic organic configurations are stable. The clumped distribution of individual organisms in morphospace reflects the existence of these underlying stable configurations.

Mayr characterises the population thinker not merely as one who stresses the uniqueness of individual organisms, but as one who is consequently driven to characterise populations of organisms in terms of their statistical features. If individuals differ, then the best one can do in trying to describe a collection of them (short of describing each one in turn) is to talk of the population average for various traits, and so forth. It is important to stress that the reliance on statistical tools for the characterisation of populations is not, in fact, the unique preserve of the population thinker. For if types are what explain concentrations of forms in morphospace, then the typological thinker, too, will need to analyse the diversity of forms found in populations in statistical terms in order to decide which types ought to be posited. If, for example, a 'type' is a stable configuration of organic matter, then one can determine which are the stable configurations by determining which areas of morphospace are most densely colonised. This will be a statistical task.

In an important article on this topic, the philosopher Elliott Sober stresses the responsibility of Darwin's cousin, Francis Galton, for developing various statistical techniques for the analysis of populations, as well as for explaining statistical properties of populations in terms of further population properties. For these reasons, Galton is Sober's prime example of a population thinker (ibid.). But Galton also believed (and Sober is aware of this) that the analysis of populations reveals the existence of underlying 'positions of organic stability' – a thoroughly typological notion. As a result of this, Galton was sceptical of the efficacy of selection to transform species. He believed we need to distinguish two senses of 'variation': 'variations proper' and 'sports'. Variations proper are minor disturbances from particular positions of stability, but because these positions are stable, organisms will generally tend to return to them, rather like the Weebles I played with as a child — these are egg-shaped plastic people which wobble when pushed but always right themselves. Sports, on the other hand, are major mutations — saltations — towards new positions of stability. Galton's view is that selection can act on variation to produce temporary change, but only saltation has the permanence to produce new species. He explains the position in his work *Finger Prints*:

> The same word 'variation' has been indiscriminately applied to two very different conceptions, which ought to be clearly distinguished: the one is that of 'sports' just alluded to, which are changes in the position of organic stability, and may, through the aid of Natural Selection, become fresh steps in the onward course of evolution; the other is that of Variation proper, which are merely strained conditions of a stable form of organisation, and not in any way an overthrow of them. Sports do not blend freely together; variations proper do so. Natural Selection acts upon variations proper, just as it does upon sports, by preserving the best to become parents, and eliminating the worst, but its action upon mere variation can, as I conceive, be of no permanent value to evolution, because there is a constant tendency in the offspring to 'regress' towards the parental type.
>
> (Galton, quoted in Provine 1971: 23)

Galton is a good example of a typological thinker in my sense of the term, and Mayr is right to say that Darwin is opposed to typological thinking of this kind. Darwin says that species are formed from natural selection acting on slight variation. His position demands, then, that these small variations, if they can be added up to produce new species, are themselves stable. But if there is a vast array of variations both slight and stable, one might share the typologist's puzzlement regarding why we see distinct species at all. Why is there any clumping in morphospace? Why, instead, don't we see organic forms smeared evenly across morphospace, reflecting the almost limitless existence of stable variation that Darwin's theory seems to demand? Darwin explicitly distances himself from a typological style of explanation, granting that while it might be appropriate in chemistry, it has no home in natural history:

> The form of a crystal is determined solely by the molecular forces, and it is not surprising that dissimilar substances should sometimes assume the same form; but with organic beings we should bear in mind that the form of each depends on an infinity of complex relations, namely on variations, due to causes far too intricate to be followed—on the nature of the variations preserved, these depending on the physical conditions, and still more on the surrounding organisms which compete with each—and lastly, on inheritance (in itself a fluctuating element) from innumerable progenitors, all of which have had their forms determined through equally complex relations.
>
> (*Descent*: 206–7)

Darwin's reading of naturalists of the typological school (people like Geoffroy St-Hilaire in France, and Richard Owen in England) pushed him to recognise the existence of an important phenomenon in nature – the clumpiness of morphospace – that required explanation. Richard Owen explained this phenomenon by positing abstract structures such as the 'vertebrate archetype' – a timeless organisational scheme, manifested in different ways in all vertebrate organisms. Darwin, in contrast, looked to explain the phenomena highlighted by typologists in a non-typological way: ' . . . why, if species are descended from other species by insensibly fine gradations, do we not everywhere see innumerable transitional forms? Why is not all nature in confusion instead of the species being, as we see them, well defined?' (*Origin*: 205).

We have already seen that Darwin looks to shared history to explain the resemblance among individuals of the same species. Darwin also invokes shared history to explain resemblances among individuals of distinct species. Darwin re-interprets Owen's archetypes as ancestors: the diverse vertebrate species appear to be variations on a common theme not because they are manifestations of a single timeless ground-plan, but because they have retained the characteristics of a common ancestor (ibid.: 416, see also Amundson 2005: chapter four). But Darwin's way of thinking about shared history does not guarantee that we should expect the world to contain species that are what he calls 'tolerably well-defined objects' (Origin: 210). If common ancestry were the only principle that we could invoke, the filled areas of morphospace would take the form of uniformly diffuse clouds with the original ancestors at their centres. There would be no clumpiness in these clouds, hence no species as we know them.

In response to this challenge, Darwin places a lot of weight on the principle of divergence of character. We should expect specialisation to be most strongly adaptive, hence we should also expect the squeezing out of intermediate forms that are jacks-of-all-trades and masters of none. In general, Darwin's reasoning is similar to that of Dobzhansky (one who, along with Mayr, pioneered the articulation of the population/typological distinction), and Dobzhansky's explicitly economic language resonates with Darwin's approach:

> The hierarchic nature of the biological classification reflects the objectively ascertainable discontinuity of adaptive niches, in other words the discontinuity of ways and means by which organisms that inhabit the world derive their livelihood from the environment.
>
> (Dobzhansky 1951: 10)

Darwin explains clumping in morphospace not by reference to universally stable types, but by reference to shifting requirements of organic environments, which demand specialisation and divergence among species.

Mayr is right, I think, to cast Darwin's innovation here as primarily philosophical in character. Specifically, Darwin breaks with typological thinking by offering a different way of thinking about the relationship between the forms that *actually* exist, the forms that are *likely* to exist given specific local circumstances, and the forms that can *possibly* exist given general laws of nature. In more philosophical terms, Darwin gives us a new way of thinking about the *modality* of species. Darwin's view makes many organic forms possible; the fact of gappiness in nature is not always to be explained by pointing to the impossibility (i.e. the instability) of forms that fall into the gaps. Rather, contingent and changeable facts – relating both to ecological demands, and to the constitution of a species – explain why many possible forms are not actual at some particular time. Were different ecological demands applied to a species at a different moment in its history quite different forms could arise instead – or at least they would not be prevented from arising by natural law.

Mayr thinks of 'population thinking' as Darwin's third great conceptual innovation after natural selection and the tree of life hypothesis. But it is hard to separate population thinking, natural selection and the tree of life. Natural selection and common ancestry are Darwin's primary resources for explaining why species are 'tolerably well-defined', hence it is primarily natural selection and common ancestry that fill the explanatory gap left by the rejection of types.

6. SPECIES NATURES

I want to close this chapter by shedding some light on the question we raised at the beginning: in what sense can a Darwinian believe in human nature, or, for that matter, apple nature or squid nature?

A strong conception of a species nature needs to go beyond the assertion that there is some set of properties shared by all species members. All tigers have mass and colour, but that is not enough to secure the existence of tiger nature, for almost all organisms have mass and colour, regardless of what species they belong to. To assert the existence of tiger nature involves a commitment to bundles of properties which are present in all tigers, and are found together only in tigers. Sets of properties like these would enable us to diagnose whether any individual is a tiger, and I will call any set that reliably enables us to do this a 'diagnostic set'. A very strong conception of a diagnostic species nature would say that for every species there is some set of properties such that every member of the species has all of them, and no member of any other species has all of them.

The Darwinian should deny that species have natures in this very strong sense. Aside from the fact that species can change markedly over time, variation introduces a new set of differences with every generation, and properties in the diagnostic set will not be immune to variation. There is, however, a fairly strong conception of a diagnostic species nature still open to us. Even if we deny that all species members share all of a set of properties peculiar to the species, we might think that all species members have some significant proportion of a set of properties, which no member of any other species has in significant proportion. This kind of piecemeal resemblance between species members would be enough to allow a reliable diagnosis of which species an individual belongs to, and it is the kind of resemblance which is compatible with the ubiquity of variation in every species.

Polymorphic forms might feature in the diagnostic set. Earlier in the chapter we saw that the males of *Paracerceis sculpta* come in three quite different sizes. We could include all three sizes in our diagnostic set. This would be especially useful for correctly identifying instances of this species if members of other, apparently similar, species are rarely found in any of these sizes. It turns out then that our conception of a species nature as a set of diagnostic properties is compatible with very large amounts of polymorphism within the species. This is important, for in species where polymorphism is ubiquitous, it will be highly misleading to talk about the species nature as something present in any individual organism. The species nature is better understood as a property of the species taken as a whole. In the case we have been considering, males of *Paracerceis sculpta* are large, small or very small, but no species member has all three sizes at once. What is more, no species member need be close to the average of the three types, just as no human has the average number of legs (which is somewhere between one and two).

Of course, when people speculate regarding the existence of human nature, they are rarely asking whether there is some set of diagnostic properties for humans. A diagnostic set could do its job even if it made reference only to properties of skulls or teeth. When people talk about human nature they are usually discussing the reality or otherwise of some set of psychological traits common to nearly all members of our species. I raise this subject only briefly here: it is discussed in detail in chapter five. For the moment we should note that polymorphism is worth remembering in this context, too, for there is no general reason to assume that selection must have made humans psychologically similar in all respects. The biologist David Sloan Wilson (1994) has suggested that the presence in human populations of both introverts and extroverts might be best understood as an instance of adaptive polymorphism. Whether Wilson is right about this depends on the facts of the case, but we should not dismiss it out of hand. If selection can maintain distinct behavioural types in a species of marine crustacean, why deny that selection could have maintained distinct psychological types in humans?

If adaptive polymorphism is widespread in humans, it will have another consequence for our conception of human nature. There is a tendency to use human nature as an explanation for individual actions: 'Why did he shout at the referee? Well, it's human nature, isn't it?' But if human nature, along with other species natures, is best understood as a property of the species taken as a whole, then human nature is not something that any one of us has, nor is it something any person could invoke to explain their actions.

SUMMARY

One of Darwin's most significant achievements is his defence of the conception of species as related in such a way that they form a vast 'Tree of Life'. But are species themselves real entities, with an existence in nature independent of the interests of human enquirers? Or do they instead reflect our practical need to catalogue and describe nature, with no independent existence of their own? In spite of Darwin's comments about the species category being one 'arbitrarily given for the sake of convenience', he shows strong inclinations to realism about species. They are, in his view, groups of resembling organisms, and these resemblances are explained by common ancestry. This makes it hard to pin either one of the dominant modern views of species on Darwin. Sometimes he seems to endorse the view that species are sets of resembling organisms, or kinds. At other times he seems to endorse the competing view of species as parts of the tree of life – that is, he regards them as 'individuals', with physical boundaries, and a beginning and end in time. Darwin's belief that new species are formed by the accumulation of slight variations leads him to the rejection of 'typological thinking' – the view that there is only a handful of stable 'types', which underlie and explain the observed clustering of organic forms. Ernst Mayr sets up 'population thinking' in opposition to typological thinking, but we must handle this distinction with care. Typological thinkers need to analyse statistical properties of populations in order to infer which types are real: in this sense, one can be a typologist and employ a form of population thinking. We might add to this that Darwin himself was poor at statistical analysis: hence it would be hard to attribute any mathematically sophisticated form of population thinking to him. Even so, Mayr is right to say that Darwin had a new way of thinking about the clustering of organic forms. Darwin explains this clustering by invoking a combination of natural selection and shared ancestry.

FURTHER READING

It is hard to pick on just a few elements of Darwin's work that relate to his views on species, but extended discussions can be found in chapters six and seven of *Descent*, and chapters two, four and especially thirteen of *Origin*.

John Dupré has written several important philosophical articles on the nature of species and taxonomy. Many of these are collected in:

Dupré, J. (2002) *Humans and Other Animals*, Oxford: Oxford University Press.

Hull and Ghiselin defend the view that species are individuals in a number of places, including:

Ghiselin, M. (1974) 'A Radical Solution to the Species Problem', *Systematic Zoology*, 23: 536–44.
Hull, D. (1978) 'A Matter of Individuality', *Philosophy of Science*, 45: 335–60.

Paul Griffiths has defended the view of species as kinds, and his view has been attacked by Mohan Matthen and Marc Ereshefsky:

Griffiths, P. (1999) 'Squaring the Circle: Natural Kinds with Historical Essences', in R. Wilson (ed.) *Species: New Interdisciplinary Essays*, Cambridge, MA: MIT Press.
Ereshefsky, M. and Matthen, M. (2005) 'Taxonomy, Polymorphism and History: An Introduction to Population Structure Theory', *Philosophy of Science*, 72: 1–21.

Griffiths's paper also discusses the topic of 'population thinking', but readers should begin with Mayr's original paper (first published in 1959, reprinted in 1976), and Sober's important reflections on the issue:

Mayr, E. (1976) 'Typological versus Population Thinking', in E. Mayr, *Evolution and the Diversity of Life*, Cambridge, MA: Harvard University Press.
Sober, E. (1980) 'Evolution, Population Thinking, and Essentialism', *Philosophy of Science*, 47: 350–83.

For a recent examination, from a philosophical perspective, of Darwinism's relationship to typological thinking, population thinking and essentialism see:

Amundson, R. (2005) *The Changing Role of the Embryo in Evolutionary Thought*, Cambridge: Cambridge University Press.

David Hull's view of species as individuals has long made him suspicious of the concept of human nature. He makes his sceptical case in:

Hull, D. (1998) 'On Human Nature', in D. Hull and M. Ruse (eds) *The Philosophy of Biology*, Oxford: Oxford University Press; originally published in A. Fine and P. Machamer (eds) *PSA Volume Two* (1986): 3–13.

Four

Evidence

I. SCIENCE AND GOD

Darwin framed the first edition of the *Origin* with two epigraphs, both from people who are probably best known now for their contributions to the study of scientific method. One is from Francis Bacon (1561–1626), a man famous for his insistence on scrupulous and exhaustive observation if scientific knowledge is to be acquired, and who, it is said, died from a cold following an experiment in which he stuffed a chicken with snow:

> To conclude, therefore, let no man out of a weak conceit of sobriety, or an ill-applied moderation, think or maintain, that a man can search too far or be too well studied in the book of God's word, or in the book of God's works; divinity or philosophy; but rather let men endeavour an endless progress or proficience in both.
>
> (Bacon: *The Advancement of Learning*)

Bacon was a highly regarded figure among many of Darwin's most influential contemporaries. They thought of him as the father of the 'inductive method' in science; that is, a method that places great store on the meticulous gathering of numerous experimental results before confidence is placed in any hypothesis. More specifically, this method warns against leaping to general theoretical positions simply because they are consonant with some small number of observations.

The 'Baconian' method is sometimes caricatured as one that exhorts the scientist to begin their work by accumulating diverse facts, thereby allowing observations to speak for themselves without the distorting bias that theoretical presuppositions might bring. Darwin claimed, rather unconvincingly, to have fashioned his theory in this way: 'I worked on true Baconian principles, and without any theory collected facts on a wholesale scale' (*Autobiography*: 72). But thoroughly Baconian science of this kind is almost impossible to conceive, and hardly seems commendable. The philosopher of science Karl Popper is particularly well-known for deriding it. He was fond of going into classrooms full of students and asking them to 'observe'. Unsurprisingly, the students were confused – how could they observe unless they had some idea of what they were supposed to be observing? Popper used this baffling request to make the point that naive observation, unclouded by any theoretical assumptions, is impossible, for we need some kind of theory to tell us where to look, and how to interpret what we see. Darwin's notebooks make it clear that he was no naive Baconian of this sort: he designed experiments in order to test the presuppositions of his transmutationist views. But it was important for him to be seen to credit Bacon, in order to signal to Bacon's Victorian admirers that Darwin's theory was built on a firm empirical foundation, and that unlike Lamarck and Chambers, he should not be viewed as one prone to wild flights of fancy.

The *Origin*'s other epigraph comes from one such influential Victorian Baconian – William Whewell:

> But with regard to the material world, we can at least go so far as this—we can perceive that events are brought about not by insulated interpositions of Divine power, exerted in each particular case, but by the establishment of general laws.
>
> (Whewell: *Bridgewater Treatise*)

As we saw in chapter one, Darwin knew Whewell from his days in Cambridge. Philosophers and historians invariably describe Whewell as a polymath: his work on the philosophy of science will be of particular concern to us here, but he also wrote on an array of subjects from astronomy to law, from architecture to mineralogy. The quotation comes from Whewell's *Bridgewater Treatise*. The *Bridgewater Treatises* were a series of works commissioned from different authors by the Earl of Bridgewater to demonstrate the existence of God, as evidenced by the natural world. Whewell's comments show that, at least as far as the physical sciences went, he did not believe that God intervenes directly to influence individual events; rather, he thought that God had set up natural laws (such as Newton's laws of motion), and those laws dictated the patterns of the Universe. As we saw in chapter one, Darwin's general view of the natural world at the time he wrote the *Origin* was of the same type. He did not believe that an intelligent God was directly responsible for creating individual species, nor for fitting species to their environments. Darwin thought that natural laws alone (primarily the law of natural selection) were responsible for these phenomena. But Darwin did not intend natural selection to rule out a God ultimately responsible for natural laws themselves.

Darwin's epigraphs raise two interrelated themes, which we will address in sequence in this chapter. First, we will look at what Darwin understood by scientific method. Specifically, we will look at what Darwin and some of his contemporaries thought it took for a theory to be well supported by evidence. We will then look at today's debate between the theory of intelligent design and the theory of evolution by natural selection. Armed with a good understanding of what it takes for a theory to have a good base of evidence, what should we make of the claim that both natural selection and intelligent design have claims to be taught alongside each other in biology classrooms? Is there really good evidence in favour of the intelligent design hypothesis?

2. INFERENCE TO THE BEST EXPLANATION

Darwin gives a strong hint in the *Origin* of his view about what makes a theory a good one. In the final chapter he summarises the varied facts his theory is able to explain – facts about anatomy, embryology, the distribution of species around the globe, even about the characteristic arguments had by natural historians – and he notes how ill-equipped are rival theories for explaining the same facts. In the *Origin's* sixth edition he adds that:

> It can hardly be supposed that a false theory would explain, in so satisfactory a manner as does the theory of natural selection, the several large classes of facts above specified. It has recently been objected that this is an unsafe method of arguing; but it is a method used in judging of the common events of life, and has often been used by the greatest natural philosophers.

> (Darwin 1959: 748)

In other words, the fact that a theory is able to successfully explain diverse phenomena is, Darwin thinks, strongly indicative of the theory's truth. This mode of reasoning has become known today as *Inference to the Best Explanation*, often abbreviated to IBE (Lipton 2004). Darwin is right that this is a highly intuitive conception of how theories of all kinds are supported by data. Why do we think it is likely that the butler killed the Earl of Wensleydale? Because if the butler did kill him, then this would best explain our data – the Earl's blood on the butler's jacket, the Earl's blood on the knife found under the butler's bed, the sighting of the butler running away from the crime scene just after the Earl died. When a hypothesis explains the data better than its rivals, we often think the hypothesis is true.

The *Origin* is full of arguments which seek to show how much better our understanding is of diverse phenomena if we assume common ancestry rather than special creation. Consider this example, where Darwin explains the different distributions of species in the Galapagos Islands (in the Pacific, nearest continent South America) and the Cape de Verde Archipelago (in the Atlantic, nearest continent Africa):

> The naturalist, looking at the inhabitants of these volcanic islands in the Pacific, distant several hundred miles from the continent, yet feels that he is standing on American land. Why should this be so? why should the species that are supposed to have been created in the Galapagos Archipelago, and nowhere else, bear so

plain a stamp of affinity to those created in America? There is nothing in the conditions of life, in the geological nature of the islands, in their height or climate, or in the proportions in which the several classes are associated together, which resembles closely the conditions of the South American coast: in fact there is considerable dissimilarity in all these respects. On the other hand, there is a considerable degree of resemblance in the volcanic nature of the soil, in climate, height, and size of the islands, between the Galapagos and Cape de Verde Archipelagos: but what an entire and absolute difference in their inhabitants! The inhabitants of the Cape de Verde Islands are related to those of Africa, like those of the Galapagos to America. I believe this grand fact can receive no sort of explanation on the ordinary view of independent creation; whereas on the view here maintained, it is obvious that the Galapagos Islands would be likely to receive colonists, whether by occasional means of transport or by formerly continuous land, from America; and the Cape de Verde Islands from Africa; and that such colonists would be liable to modifications;—the principle of inheritance still betraying their original birth place.

(*Origin*: 385–86)

The hypothesis that an intelligent God produced each species individually to suit its surroundings leads us to expect that similar habitats will contain similar species, and different habitats will contain different species. This is not what we see. Compared with Darwin's hypothesis of common ancestry, special creation is a poor explanation of our data, and using IBE it is far less likely to be true.

We are undoubtedly prone to think theories true when they have explanatory power. But some explanations have lots of attractive attributes – they tie disparate phenomena together, they show us that the events we have witnessed are ones we should have anticipated – yet they are, in spite of all this, false. Conspiracy theories are often like this. So although 'Inference to the Best Explanation' is an attractive slogan, ideally one would want to flesh it out with an account of what makes an explanation good, and an account of why, when a hypothesis ties up some body of facts in a neat, explanatory way, it is therefore likely to be true. One cannot solve the first problem by insisting that only true explanations are good ones. IBE's defender needs to pinpoint characteristics of good explanations, and then show that explanations with these characteristics are more likely to be true than bad ones. Requiring that an explanation does not count as good unless it is true misses the point of IBE.

As a first pass, we might understand a good explanation to be one that cites factors which would raise the probability of the event we are trying to explain. The more the probability is raised, the more satisfying the explanation (Mellor 1976). The most satisfying explanations are the ones that show that what happened had to happen, with 100% probability. On this view, we do not define a good explanation as one that is true, we define it as one that, were it true, would make the facts we seek to understand probable. We can explain a fire by reference to a short-circuit because in the circumstances (a ready supply of oxygen, a warehouse without sprinklers), a short circuit would make a fire highly probable.

When a hypothesis makes some set of observations very probable, philosophers and statisticians say that the hypothesis has a high *likelihood*. I will use italics throughout this chapter to refer to this technical notion of *likelihood*. It is important to remember that the *likelihood* of a hypothesis is a function of how probable the hypothesis makes some set of observations, not of how probable the observations make the hypothesis. So, for example, given the observation that there is clear liquid dripping down my kitchen walls, the hypothesis that there is a flooded bath upstairs has high *likelihood*. This hypothesis makes my observation likely. The proposal we have been considering is that good explanations of observations are hypotheses with high *likelihoods* in the light of those observations. Are explanations that are good in this sense also generally true? They are not, as the philosopher Elliott Sober often reminds us (e.g. Sober 1993). For any given set of data, there are typically numerous alternative hypotheses that make those data probable. They all have high *likelihoods*, but we must choose which is true. The drips on my walls are made likely by the flooding bath, but could also have been made by spilled vodka, errant guttering, a burst pipe and so forth. Some hypotheses with high likelihoods are patently absurd. If I hear a drumming noise coming from my ceiling, then I can explain this either by hypothesising that it is raining, or by hypothesising that tiny fairies are having a disco on the roof. Both hypotheses make my observations probable – if there were a fairy disco on the roof, then I would hear a drumming noise – but I do not regard the fairy hypothesis as probably true in virtue of this, even though that hypothesis has a high likelihood, and is, by the criterion we are currently discussing, a good explanation.

This teaches us an important lesson. We should not accept hypotheses merely on the grounds that they make some body of data highly probable. This does not mean that the likelihood of a hypothesis is irrelevant to its truth, but it does show that likelihood is not sufficient to justify belief in a hypothesis.

3. HERSCHEL AND WHEWELL

The methodologists of science who influenced Darwin most heavily – especially John Herschel – insisted that scientific theories should appeal only to what they called *verae causae*, or 'true causes'. Darwin read Herschel's major methodological work – *A Preliminary Discourse on the Study of Natural Philosophy* – during his student days:

> During my last year at Cambridge I read with care and profound interest Humboldt's *Personal Narrative*. This work and Sir J. Herschel's *Introduction to the Study of Natural Philosophy* stirred up in me a burning zeal to add even the most humble contribution to the noble structure of natural science. No one or a dozen other books influenced me nearly so much as these two.
>
> (*Autobiography*: 36)

We can understand adherence to the *vera causa* ideal as a recognition of the point we made in the last section. High likelihood does not by itself constitute serious evidence in favour of a theory. A theory must do more than explain some narrow class of phenomena if it is to command our confidence. There is no evidence in favour of the fairy hypothesis beyond the fact that it fits with our observation of noise from the roof. Rain, on the other hand, is something we have plenty of additional evidence for. 'True causes' are causes like the rain, rather than dancing fairies.

What, precisely, does it take for us to have 'additional evidence', of a kind that might begin to warrant our believing an explanatory theory? Equivalently, we can ask what it takes for a theory to meet the *vera causa* ideal (for details see Ruse 1975; Hodge 1977). Herschel gives various different answers to this question in his *Preliminary Discourse on the Study of Natural Philosophy*. At some points he adopts a proposal consonant with a more plausible version of IBE: *verae causae* are those which, in addition to explaining the phenomena that initially lead to their invocation, are also able to explain many other phenomena. They are causes 'competent, under different modifications, to the production of a great multitude of effects, besides those which originally led to a knowledge of them' (Herschel 1996: 144). Maybe we should believe in fairy discos if such discos also explained phenomena beyond the drumming noise on the roof (if, for example, they explained the appearance of tiny ethereal beer cans, left discarded on the mornings after drumming is heard). At this point, Herschel says, the scientist can be fairly certain that he has found 'causes recognized as having a real existence in nature, and not being mere hypotheses or figments of the mind' (ibid.). Best of all is when, once we have posited some cause, we are then able to form further explanations we had not expected, as well as explanations of phenomena that, at first glance, seem opposed to our theory:

> The surest and best characteristic of a well-founded and exten-
> sive induction, however, is when verifications of it spring up, as it
> were, spontaneously, into notice, from quarters where they might
> be least expected, or even among instances of that very kind
> which were at first considered hostile to them. Evidence of this

kind is irresistible, and compels assent with a weight which scarcely any other possesses.

(Ibid.: 170)

In these respects, Herschel is close in his methodological stance to Whewell, who claims in his *Philosophy of the Inductive Sciences* that:

. . . the evidence in favour of our induction is of a much higher and more forcible character when it enables us to explain and determine cases of a *kind different* from those which were contemplated in the formation of our hypothesis. The instances in which this has occurred, indeed, impress us with a conviction that the truth of our hypothesis is certain.

(Whewell 1996: 230, emphasis in original)

Both men recommend confidence in hypotheses that explain many diverse phenomena. When a hypothesis does this we have what Whewell calls the *Consilience of Inductions*. One difference between the two men, and it is a mild one, is that while Herschel puts the stress on explanations of phenomena in domains initially thought hostile to the theory, Whewell stresses explanations of kinds of phenomena which the theory was not designed to accommodate (Laudan 1981).

At times Herschel makes an even stronger demand on a theory, namely that we should be able to *directly perceive* either the causes it refers to, or something closely analogous to them. At this point, Herschel's demands go well beyond Whewell's, because a cause that was unlike anything humans could ever perceive could still explain diverse phenomena, including phenomena theretofore unanticipated. It is tempting to say that Whewell, not Herschel, got it right here, for Herschel's strong requirement would seem to disqualify successful theories in fundamental physics, which posit causes wholly alien to anything we might perceive. This conclusion is a little too hasty, however, for even in more recent years some philosophers have argued that a scientific theory cannot be understood, let alone believed, unless its claims can somehow be linked, by analogy, with systems we are familiar with (e.g. Hesse 1966).

Herschel insisted on the requirement of direct perception because he regarded Newton's explanation of the orbits of the planets as a paragon of solid science:

> For instance, when we see a stone whirled round in a sling, describing a circular orbit round the hand, keeping the string stretched, and flying away the moment it breaks, we never hesitate to regard it as retained in its orbit by the tension of the string, that is, by *a force* directed to the centre; for we feel that we do really exert such a force. We have here *the direct perception* of the cause. When, therefore, we see a great body like the moon circulating round the earth and not flying off, we cannot help believing it to be prevented from so doing, not indeed by a material tie, but by that which operates in the other case through the intermedium of the string,—a *force* directed constantly to the centre.
>
> (Herschel 1996: 149, emphasis in original)

Herschel believed that the key to Newton's success lay in his appeal to a notion of gravitational force whose existence, and whose ability to explain orbital motion, were already experienced in analogous form by human observers.

4. HERSCHEL AND THE *ORIGIN*

Let us spell out Herschel's view in more detail. He thinks a theory should command our assent only when the causes it appeals to pass three tests. They must be: (1) shown to *exist*; (2) shown to be *capable* of producing the phenomena we seek to account for; and (3) shown to be *responsible* for producing the phenomena we seek to account for (Hodge 1977). Fairies fail the first test of existence, because beyond the fact that fairies might explain the drumming noise on the roof, we have no reason to think there are fairies. Do fairies pass the capability test? That is not clear, for it is hard to decide whether they could produce a drumming noise unless we have information about what kinds of things fairies are. That we have no such information is partly a symptom of their failure to pass the existence test. This shows that in some cases it is hard to pull the existence test and the capability test apart. Many competing hypotheses will pass both of these tests: the drumming noise on the ceiling might be produced by rain, or by hailstones, and both of these factors we know to exist and to be capable of producing a drumming noise. One establishes that rain, rather than hail, is responsible for the noise on a particular occasion (one shows it alone passes the third test) by cataloguing further phenomena that rain is better able to explain than hail (the flow of water down gutters, for example).

Darwin's structuring of the *Origin* seems to answer each of Herschel's three requirements in turn, as the historian Jonathan Hodge has argued (ibid.). First, Darwin tries to demonstrate that natural selection exists, and that it is capable of producing new species and adaptation. He does this in a way that very closely parallels Herschel's reconstruction of the proper Newtonian method (Ruse 1975). Whirling a stone in a sling gives us direct perception of a force capable of producing circular motion. This, says Herschel, should increase our confidence in the existence of an analogous force exerted by the Earth on the Moon. Darwin tries to establish natural selection as a *vera causa* in a similar way, by appealing to our direct experience of an analogous force in artificial selection.

As we saw in chapter two, Darwin uses the success of animal breeders to make his case for the existence of an abundant supply of natural variation that is not under human control, and which provides raw materials suitable for adaptive modification. If natural selection is to produce adaptation in the wild, Darwin also needs to show that these variations can be preserved and added up without the conscious hand of a breeder. Darwin uses artificial selection to establish this claim, too.

First, he points to cases where humans, as part of the natural environment of a species, cause considerable adaptive modification merely by looking after their most valuable animals, rather than by intentionally instituting a breeding programme to modify them:

> If there exist savages so barbarous as never to think of the inher-
> ited character of the offspring of their domestic animals, yet any

> one animal particularly useful to them, for any special purpose,
> would be carefully preserved during famines and other accidents,
> to which savages are so liable, and such choice animals would
> thus generally leave more offspring than the inferior ones; so that
> in this case there would be a kind of unconscious selection going on.
>
> (*Origin*: 94)

To say that Darwin makes a case for the existence and capability of natural selection by drawing an *analogy* with artificial selection is slightly misleading here; rather, Darwin shows that some cases of artificial selection are cases of natural selection.

Second, Darwin tries to establish that natural selection is capable of producing complex adaptations, by suggesting that natural selection is far more powerful than artificial selection:

> As man can produce and certainly has produced a great result by
> his methodical and unconscious means of selection, what may
> nature not effect? Man can only act on external and visible char-
> acters: nature cares nothing for appearances, except in so far as
> they may be useful to any being. She can act on every internal
> organ, on every shade of constitutional difference, on the whole
> machinery of life . . . How fleeting are the wishes and efforts of
> man! how short his time! and consequently how poor will his
> products be, compared with those accumulated by nature during
> whole geological periods.
>
> (ibid.: 132–33)

In brief, we can see the argument of the *Origin* as a Herschellian one. The case relating to the existence of natural selection and its capability of producing new species and adaptation takes up the first eight chapters of the *Origin*. Here Darwin is particularly keen to address, for example, those phenomena that one might think his theory could not possibly account for, such as 'organs of extreme perfection' (exquisitely designed adaptations like the eye), behaviours that appear to be of no value to the individual organism, and so forth. Chapters nine to thirteen then seek to show that natural selection is not merely capable of effecting organic change, but is actually responsible for it, by showing how Darwin's theory makes far better sense of diverse phenomena than do competing theories, especially the theory of special creation.

5. DARWIN, WHEWELL AND GEMMULES

The case for reading the *Origin* as a long Herschellian argument is, I think, very strong. But in later works Darwin appears to toy with relaxing Herschel's strong standards, and moving more closely in line with Whewell (Ruse 2000a). This is particularly true of the case that Darwin builds for his theory of inheritance – *pangenesis* – which he defends in *The Variation of Animals and Plants under Domestication*. In the *Origin*, the fact that offspring generally resemble their parents is taken largely for granted, and Darwin simply admits his ignorance regarding why this should be the case. Darwin was not content with this, and sought to put forward a theory that would account for these resemblances. He believed this theory should be able to explain the recognised phenomena of inheritance. So, for example, the theory should be able to explain not only why offspring resemble parents, but why traits sometimes 'jump' a generation, appearing in grandchildren and not children. Darwin also believed that alterations to a parent during his or her lifetime could sometimes appear in children. His theory should be able to explain this, too. Finally, the theory should be able to explain the special role of the sex cells (sperm and eggs) in inheritance.

Darwin argued that each part of the body produced particles, called 'gemmules', which were of a character specific to that part. He thought that gemmules specific to each part of the body gathered in the sex cells, with the result that an embryo formed from the union of sperm and egg would contain a full set of different types of gemmules from each parent. Once an embryo is formed, gemmules can either develop to produce traits resembling those which produced them, or they can remain dormant for several generations. In these last respects gemmules are not entirely unlike genes. But Darwin's theory sought to explain many phenomena whose reality we no longer acknowledge. Darwin proposed that gemmules migrated from organs to the sex cells in order to explain how alterations to an organ during the life of an individual could result in alterations to the gemmules in that organ, ultimately changing the character of the gemmules transmitted to future generations. This was how the blacksmith's son would come to inherit the smith's biceps. The problem, of course, is that we no longer believe that acquired variations of these kinds are passed on to offspring.

Darwin's Herschellian instincts show themselves in his description of the hypothesis of pangenesis as 'provisional'. He was reluctant to claim a firm conviction in his ideas, for no one had directly perceived gemmules, or anything analogous to them. On the other hand, Darwin begins to wonder if Herschel's demand for direct perception of one's explanatory posits may be too strong. He suggests that the explanatory power of a theory alone may be justification enough to warrant belief.

Most of Darwin's contemporaries accepted the view that light was a wave, travelling via undulations in an invisible medium called the luminiferous ether. The luminiferous ether was, in principle, imperceptible. In one of Darwin's longer reflections on method, he flirts with a Whewellian consilience of inductions as the evidential foundation for the natural selection theory:

> In scientific investigations it is permitted to invent any hypothesis, and if it explains various large and independent classes of facts it rises to the rank of a well-grounded theory. The undulations of the ether and even its existence are hypothetical, yet every one now admits the undulatory theory of light. The principle of natural selection may be looked at as a mere hypothesis, but rendered in some degree probable by what we already know of the variability of organic beings in a state of nature,—by what we positively know of the struggle for existence, and the consequent almost inevitable preservation of favourable variations,—and from the analogical formation of domestic races. Now this hypothesis may be tested,—and this seems to me the only fair and legitimate manner of considering the whole question,—by trying whether it explains several large and independent classes of facts; such as

the geological succession in organic beings, their distribution in past and present times, and their mutual affinities and homologies. If the principle of natural selection does explain these and other large bodies of facts, it ought to be received.

(*The Variation of Animals and Plants under Domestication,* quoted in Gayon 1998: 32)

Darwin asserts, in conformity with Whewell's notion of consilience, that natural selection's explanation of not just one, but 'several large and independent classes of facts', counts strongly in its favour. But a quasi-Herschellian addition to Darwin's case remains: that natural selection exists as a force competent to produce adaptation and speciation is 'rendered in some degree probable' by the fact that species vary, and by the fact that conditions in nature are demanding (Hodge 2000).

Let us step back from the case of Darwin, and ask what we should make of the dispute between Herschel and Whewell. Should we insist on the 'direct perception' of the causes posited by some theory, or is explanatory power enough? One problem is that it is hard to know what 'direct perception' means. Do I directly perceive the postman when I hear him knock on the door? Do I directly perceive atomic nuclei on the screen of an electron microscope? At the same time, it is not clear why 'direct perception' should be important for confirmation of a theory. There are many cases where we claim to have perceived something directly that is in fact a figment of the imagination. This is what happens when we experience vivid hallucinations. It is no use replying that direct perception is important because one cannot perceive something unless it exists: this simply transforms the problem into how to tell that you really are perceiving something when it seems that you are.

The work of the philosopher Ian Hacking contains resources that might help solve both problems at once. He describes a striking encounter with modern scientists who do not merely talk of their good reasons for believing in subatomic particles, but casually mention the uses to which they put those particles. How do they change the charge on a microscopic ball of niobium? By spraying it with positrons (Hacking 1983: 23). These scientists do not only posit positrons in order to make sense of diverse phenomena, they take it that they can use and manipulate positrons. When an entity is thought to be 'manipulable' in this way, our confidence in its existence is certainly increased. Whether we 'directly perceive' positrons is moot, but Herschel is clearly onto something when he claims that the strongest evidence in favour of a theory comes from intimate familiarity with the causes it posits, a familiarity that we can perhaps understand through Hacking's concept of manipulability.

On close inspection, it may be that Whewell can incorporate the special significance of manipulation into his concept of consilience. To appear able to manipulate an entity is to act in various ways as though the entity exists, and for one's actions to have the results predicted by the alleged properties of that entity. Successful manipulation thus consists in a series of events of different kinds (different types of interactions and their results), which the theoretical properties of the posited entity are jointly able to explain. The proponent of consilience can understand manipulation in terms of the ability of a theory of what a scientific entity is like to explain a series of events associated with our interventions and their consequences. So Whewell, too, can explain why manipulation is so important for the confirmation of a theory, by reference to his idea that a good theory is one that explains a diverse set of phenomena. Darwin may be right to keep his feet in both Herschellian and Whewellian camps, for it is not clear that we need to choose between them.

6. NATURAL SELECTION AND COMMON ANCESTRY

As I have emphasised, it is one thing to claim that species are descended from common ancestors to form a genealogical tree, another thing to say that natural selection is the specific process responsible for this state of affairs. When Darwin talks about 'my theory', even when he talks about 'the principle of natural selection', he is often ambiguously situated between these two claims. So when Darwin tells us that 'his theory' is able to successfully explain a host of diverse phenomena in consilient fashion, is he talking about common ancestry, natural selection, or both?

The early parts of the Origin focus on establishing that natural selection exists, and that it is capable of producing adaptation and speciation. But the philosopher Kenneth Waters points out that many of the phenomena which, in the second half of the Origin, Darwin argues are best explained in terms of 'his theory', are explained in terms of common ancestry alone (Waters 2003). It is common ancestry that explains why species are similar in the Galapagos and the American mainland, yet different in the Galapagos and the Cape Verde Islands. The same is true for many of the other sets of facts Darwin accounts for, such as the strong resemblances we see between organs with different functions:

> What can be more curious than that the hand of a man, formed for grasping, that of a mole for digging, the leg of the horse, the paddle of the porpoise, and the wing of the bat, should all be constructed on the same pattern, and should include the same bones, in the same relative positions? . . . On the ordinary view of the independent creation of each being, we can only say that so it is;—that it has so pleased the Creator to construct each animal and plant.
>
> The explanation is manifest on the theory of the natural selection of successive slight modifications,—each modification being profitable in some way to the modified form, but often affecting by correlation of growth other parts of the organisation.
>
> (*Origin*: 415–16)

The arguments Darwin mounts in the Origin leave special creation – the view that species are individually created by an intelligent agent – dead and buried. It is baffling why an intelligent God should choose to build the horse, the mole and the bat on the same anatomical plan when their environments make such different demands on them. But although Darwin tells us in this passage that it is the theory 'of the natural selection of successive slight modifications' that explains these observations, it is not clear from what he says that these observations can only be explained by the specific mechanism of natural selection. Might other transmutationist theories − such as Lamarck's, or his grandfather's − also be able to accommodate the phenomena Darwin describes here? We saw in chapter two that Darwin failed to address some 'populational' worries about the capability of natural selection to produce adaptation. When we add to this the fact that in the second half of the *Origin* Darwin is mainly concerned with exposing the explanatory inadequacies of special creation, rather than natural selection's specific strengths, we understand better why almost all of Darwin's contemporaries were quick to accept his case for evolution, but fewer were moved by the argument for natural selection.

7. THE NATURAL SELECTION/INTELLIGENT DESIGN DEBATE

A group of vocal and politically influential critics of Darwin continues to argue not merely that natural selection remains in doubt as the means by which species have become adapted to their environments, but that intelligent design offers us a better hypothesis to account for adaptation (e.g. Behe 1996; Dembski 2004). They also use IBE to defend their view. We are now in a good position to examine their argument, and to expose its failings.

The biochemist Michael Behe uses a structure present in bacteria called the flagellum as the basis for the case for intelligent design (Behe 1996). The flagellum is a whip-like filament that twirls around, thereby propelling the bacterium. Behe thinks the flagellum is better explained by intelligent design than by natural selection, and as a result he argues that the intelligent design hypothesis is likely to be true. He does not claim that the flagellum will show all by itself that a god of the kind Christians believe in exists, but he does think the flagellum points to a designer of some kind.

Here are the four main steps in the intelligent design argument, as it is usually phrased today. In this section, I will assess each one in turn.

I)   Natural selection cannot explain the flagellum
II)   Chance is a poor explanation of the structure of the flagellum
III) The intelligent design hypothesis is a good explanation of the
      structure of the flagellum
IV) IBE demands assent to the intelligent design hypothesis.

I) Natural selection cannot explain the flagellum

Behe thinks the flagellum manifests something he calls 'irreducible
complexity'. He says that if any part of the flagellum were removed
it would be unable to perform any function. How, then, can we
think that the flagellum is produced by numerous small mutations,
each one fitter than the last? Behe's objection is similar to one
levelled in Darwin's time by St George Mivart (1871). What use,
said Mivart, is half an eye? Similarly, says Behe, a partial flagellum
is good for nothing, and he concludes that the flagellum could not
possibly have evolved through natural selection acting on gradual
mutation.

A good response to Mivart is to point out that half an eye might
be very useful indeed, especially when other members of the species
have no eyes at all (Dawkins 1986). Similarly, several biologists have
sketched possible evolutionary histories for the flagellum, which
credit partial flagella with various useful functions (Young and Edis
2004). These partial flagella do not have the same functions as the
full flagellum – many biologists think the flagellum is descended
from an organ used for secretion, rather than propulsion – but selection explanations require only that each successive modification is
fitter than the last, not that it is fitter for the same reasons. (Similarly,
insect wings may first have evolved as devices for regulating temperature. Stubby protuberances on the sides of the body do not assist
in flight all that much, but they do help with cooling [Kingsolver
and Koehl 1985].) To cite a possible history for the flagellum is not
to establish how it did in fact evolve, but remember that Behe is
not claiming merely that we are unsure about how the flagellum
evolved, he is claiming that it could not have evolved.

Three further points are worth making here. First, intelligent
design theorists sometimes try to show how unlikely it is that

selection would produce something as elegant as the flagellum by calculating the tiny chances of all the parts of the flagellum being thrown together at random in the proper conformation. This calculation is irrelevant. As we saw in chapter two, adaptations can be far more likely to be produced by natural selection than they would be from a random coming-together of matter.

Second, suppose Behe turns out to be right that if any part of the flagellum were removed, the remaining structure would give no advantage to survival and reproduction in any respect. It does not follow that natural selection cannot explain the flagellum. That is because the flagella in the ancestors of today's bacteria might have contained many more parts than the flagella we see now. They may have been rather over-complex, ramshackle structures, which have been improved by the removal of parts, to leave a pared-down structure from which no more parts can be removed without loss of functionality. And these over-complex, ramshackle structures could, perhaps, have been built by gradual steps from earlier ancestors, for they are not so delicately organised that all functioning is lost if one element is absent.

Third, suppose that things turn out badly for selection's advocates, and we really can find no plausible selection explanation for the flagellum's structure. The most we could conclude from this is that we don't know how to explain it. This would hardly show all by itself that an intelligent agent probably is responsible, any more than scepticism about whether rain is the cause of the drumming on my ceiling should lead me to expect that fairies are most likely responsible instead (Sober 2004).

II) Chance is a poor explanation of the structure of the flagellum

If things go badly for selection, could we explain the flagellum by appealing to chance instead? Might we simply shrug our shoulders and say, 'Yes, the construction of the flagellum is exceptionally unlikely, but unlikely things can happen all the same'? Intelligent design theorists are right to say that explanations like this one are poor. But why suppose that we are entitled to a satisfying explanation of every fact? Surely sometimes very unlikely things do happen, and there is not much we can say about them other than to concede that they have happened.

Here is where intelligent design theorists make two grave slips. First, they assume that everything must be explained, and second, as a corollary of this, they assume that once we reach a certain threshold of probability, the claim that something happened by chance is unacceptable. This is a key move in the efforts of another leading intelligent design theorist, William Dembski, to formalise the inferences we make when we conclude that a structure has been produced by design (Dembski 1998). Dembski does not mind accepting that some unlikely things happen, but not events that are both really, *really* unlikely, and 'specified'.

Dembski gives a very technical definition of specification, which I will not discuss here. So far as I can see, he is trying to capture the idea that while many undesigned objects – stony beaches, forest floors – have very complicated or unlikely structures, designed objects – cars, houses – tend to have structures which, in addition to being complicated, are also well-suited to discharging useful purposes. In Dembski's language, a designed object has the sorts of features that might be identified, or 'specified', prior to the observation of the object in question. These might include design features that discharge the functions which likely users care about. For example, one does not need to have seen a car to know that an object that helps people to get around quickly is something that a person might aim to construct. Contrast this with the precise arrangement of the rocks on the beaches in Cornwall. This arrangement would be excellently suited to the purposes of someone who likes rocks in a particular pattern – namely, just the pattern they have fallen into – but we do not conclude that this arrangement was designed.

The obvious explanation for why we infer design when presented with a car, but not when presented with a Cornish beach, is that while we know that people aim to get around, and that they have the power to make medium-sized metal objects, we have no reason to think that anyone has the ability or the inclination to arrange a whole beachful of rocks. Dembski, however, gives a rather different analysis. His view is that in the case of the beach, our ability to spell out the pattern desired by our hypothetical beach-arranger relies on our knowledge of how the rocks have actually fallen. This pattern, therefore, is not 'specified'. Conversely, 'the actualization of a possibility . . . is specified if the possibility's actualization is independently identifiable by means of a pattern' (Dembski 2001: 562). Dembski appears to be saying that what matters is whether our ability to describe the pattern the rocks have adopted is independent of the fact that they have already ended up in that arrangement.

Taken at face value, this condition seems far too liberal. For it is possible (albeit tedious and time-consuming) to describe the pattern the rocks on the beach have adopted without observing them. One need only describe all the possible patterns they might fall into, and the actual pattern will be among these. The pattern of rocks on the beach is therefore specified, and it is also unlikely. Yet we should not take this to mean that the Cornish beaches have been ordered in minute detail by an intelligent designer.

Let us be more charitable to Dembski. At the very least it seems that he must count an event as specified when it is functional for some actual agent, or group of agents. This version of the design inference tells us that if some state of affairs is well-suited to a beneficial end, and if that state of affairs would be exceptionally unlikely to have arisen by chance, then it didn't happen by chance, but by design. How unlikely does a specified event have to be for us to infer that it happened because of design? This is given by what Dembski calls the 'universal probability bound', which he estimates as 1 in 10^{150}. Specified events that are this unlikely, Dembski says, just don't happen (Dembski 2004).

Dembski's design argument does not work. Consider two competing hypotheses, both of which might explain the fact that Sam wins the National Lottery. First, Sam wins the lottery fair and square; second, Sam wins the lottery by fixing the machines to yield his numbers. The fact of Sam's winning is made likely by the second hypothesis, it is not made likely by the first. Sam, like all of us, has a very small probability of winning the lottery by chance. Should we conclude that Sam probably did not win the lottery fairly after all? Is this simply too unlikely to happen by chance? No: it is an unlikely event, and a 'specified' one (it is certainly good for Sam that he wins), which we have reason to believe occurred.

One might reply that so long as many millions of people bought lottery tickets it was highly likely after all that someone would win. That is true, but it is not relevant to the claim that unlikely things happen. It merely changes the subject (Sober 2004). Showing that *someone* was likely to win does not raise the chances of *Sam* winning; *that* fact remains very unlikely on the hypothesis that the lottery is fair, but we do not dismiss it because of that.

One might reply instead that the probability of Sam's winning the lottery is indeed low on the hypothesis that the lottery was fair, but it does not approach the tiny chances of a flagellum being created through a random series of events, and it does not get close to Dembski's universal probability bound of 1 in 10^{150}. Perhaps not – the chances of winning the jackpot on the UK National Lottery are a practical certainty in comparison at 1 in 13,983,816 – but there are series of events whose existence we do not deny, and whose chances are far lower than Sam's chances of winning the lottery. Suppose André wins the lottery the week after Sam does, and Bob wins the next week. The probability of Sam and André and Bob all being lottery winners is much lower than the probability of Sam winning the lottery. If we assume that the chances of each one winning are 1 in 10^7 (i.e. one in ten million), then the chances of all three winning are 1 in 10^{21}. By adding yet more winners we can generate even lower probabilities until eventually we meet the universal probability bound. But we do not conclude that all of the actual lottery winners there have been are probably members of a powerful lottery-fixing cartel, nor do we conclude that they all won because God intended that it be so. Sometimes, very unlikely things happen. That includes very unlikely things that are functional, such as winning the lottery. The argument from organic adaptation to intelligent design is no better than the argument from lottery winners to intelligent design.

III) The intelligent design hypothesis is a good explanation of the structure of the flagellum

Does the hypothesis of intelligent design explain the structure of the flagellum? Let us return to our conception of a good explanation as one that raises the probability of the phenomena we are trying to explain. The information that a guiding intelligence is at work influencing the natural world does not of itself make structures like the flagellum probable – why suppose that an intelligent agent would make a twirly bit of bacterium, instead of somewhere to live, or a nice garden to spend time in, or someone to watch TV with? Taken on its own, the hypothesis that there is an intelligent designer is a bad explanation of the flagellum.

Any hypothesis that makes the structure of the flagellum probable, and thereby explains that structure, needs to make reference not just to the intelligence of a designing agent, but to the goals and powers of that agent (ibid.). Take the hypothesis that there is a powerful agent who has a penchant for whip-like structures, who wishes to ensure that bacteria can get around well, and who, somehow, has the means to bring these desires about. This hypothesis would, if true, make the structure of the flagellum reasonably likely, and as such it is a good explanation.

IV) IBE demands assent to the intelligent design hypothesis

Of course, showing that there is a good intelligent design explanation for the structure of the flagellum does not show that we should believe that explanation to be true. This is precisely the point that we stressed at the beginning of the chapter, when we showed that likelihood was not enough for us to accept a hypothesis. If IBE tells us to believe any hypothesis that would, if true, explain the data we have, then IBE does indeed tell us to believe that there is a designer set on equipping bacteria with flagella. IBE, understood in this way, also tells us that there is a designer who is keen on arrangements of rocks with the exact pattern of those on the beaches in Cornwall, because this hypothesis makes Cornish rocks' falling into that precise pattern very likely. IBE, understood in this way, is absurd.

As we saw when we discussed Herschel and Whewell, the explanatory hypotheses that we should believe must do far more than make some narrow class of facts probable. We do not assume that Sam, who wins the lottery, also fixed it, merely on the grounds that if he had fixed it this would make his winning more likely. Some kind of additional evidential support is needed before we put trust in an explanatory hypothesis.

Intelligent design performs poorly on all of the most plausible accounts of how a theory can gain the required evidential support. We might begin to take the lottery fixing hypothesis seriously if we could show that Sam had the desire and the ability to rig the machines. We might do this by looking at his whereabouts on the run-up to Lottery Night, or by investigating his connections with the lottery management company. Intelligent design theorists do not offer analogous evidence for the desires and abilities of the hypothesised designer. They do not investigate the whereabouts of the designer at the time that flagella are alleged to have been created, nor do they look at whether the designer was in a position to influence the molecules that make up this structure. They merely repeat their assertion that if there were a designer with the right desires and abilities, such an agent would be able to explain the flagellum.

What about the idea, backed by Herschel and Whewell, that an explanatory theory is trustworthy when it explains more than one set of phenomena? Darwin's theory begins by positing natural selection as the explanation of good adaptation, but this theory can also immediately explain poor and arbitrary adaptive solutions. The male genito-urinary system offers two examples of solutions of this sort. First, the urethra of human males passes (needlessly) through the prostate gland on its way from the bladder to the penis. A consequence of this is that as the prostate grows with age, it constricts the urethra, making urination difficult. Second, the testes in human males are linked to the penis via long tubes which wind an unnecessarily tortuous route up over the ureters (the tubes which convey fluid from the kidneys to the bladder) and back down again to meet the prostate gland (Williams 1996: 142). Poor or ramshackle adaptations of this sort are exactly what we should expect to see if organic traits are produced by gradual modifications from earlier ancestors. Intelligent design runs into trouble here, for if the flagellum is supposed to necessitate an extremely powerful designer for its production, it then becomes unclear why such a powerful designer would also produce the many botched jobs and cobbled-together solutions we see in the natural world.

We might save intelligent design by reformulating the theory. Why insist that the designer produces only good designs? Why not say that there is an intelligent designer, who wishes to make the organic world just as we find it, and who has whatever powers are required to achieve this? This new theory explains all the observations we make. Poor design is no problem, because the designer intends some design to be poor. One thus formulates a 'theory of everything' while never leaving one's arm-chair: 'Everything is as the Intelligent Designer intended it'. But this theory is equivalent to a series of isolated, and unconnected, hypotheses linking features of the supposed designer to features of the world. Why did Sam win the lottery? Because a designer intended Sam to win the lottery, and is able to make him win. Why do bacteria have flagella? Because a designer intended bacteria to have flagella, and is able equip them with flagella. We have no good reason to accept any of these claims, for they only explain the isolated facts they are invoked to account for.

8. EVOLUTION WITH INTELLIGENT DESIGN

The explanatory hypotheses we put most confidence in often make predictions that are subsequently confirmed. The more specific those predictions are, the better for the hypothesis if they are observed. Sam wins the lottery. If he were to have fixed it, then as well as winning, we might also expect that he would have tampered with the machines in some way. If subsequent investigation shows Sam's fingerprints to be on the lottery machines, then we should increase our confidence in the fixing hypothesis.

It is easy to point to successful predictions of Darwin's theory. If Darwin is right about the Tree of Life, then we should expect to find fossil remains of species that are intermediate in form between the species we see today, and we should expect to find them in rocks of appropriate dates. This is exactly what we do find: 'missing links' are being discovered all the time. For example, in the week that I was making final revisions to this chapter, the journal *Nature* reported the discovery of fossilised remains of a crocodile-like fish, named *Tiktaalik roseae*, which lived 375 million years ago (Daeschler *et al.* 2006). *Tiktaalik* marks the transition of animal life from water to the land: its skeleton is intermediate in form between fish with fins and vertebrates with limbs, and is suggestive of life in shallow water. But how far does the discovery of missing links of this sort count against intelligent design? The problem is that the intelligent design hypothesis tells us that intelligence is responsible for complex adaptation; it does not tell us that species were specially created. Some intelligent design theorists today – most notably Michael Behe – do not deny evolution. Their position is compatible with a kind of hybrid view: species are indeed descended from common ancestors, but intelligent design is responsible for complex adaptations in those species. The predictive successes of the Tree of Life hypothesis do not count against this form of intelligent design, for this form of intelligent design is compatible with the Tree of Life.

Still, let us try to ascertain what the intelligent design hypothesis does predict. One problem comes from the fact that we cannot say what kind of structures a designer will produce unless we have some conception of what the designer cares about, and how competent he or she is. Yet intelligent design's proponents almost never stick their necks out regarding the character of the designer. A second problem derives from the peculiarity of the hypothesis we are being asked to consider. There is no logical inconsistency in the view that although species are genealogically related, their adaptations owe their structure to intelligent oversight. Indeed, this is true of many species that owe parts of their structure to human breeders (think of the udders of Friesian cows, for example). But surely the intelligent design hypothesis is not that there is some invisible breeder in the sky, who has modified organic lineages through artificial selection? After all, if the intelligent design theorists are right that natural selection cannot account for the flagellum, then the standard operation of artificial selection cannot account for it either. Whatever powers the intelligent designer has, they are not supposed to be constrained in the ways that natural selection is constrained. But how are they constrained? Unless this is spelled out, we do not know what predictions the intelligent design hypothesis makes.

Perhaps we can derive some predictions from the intelligent design hypothesis by drawing on our knowledge of how intelligent design generally works. When we are designing systems, we do so part by part. Instead of building a new device for the production of rotary motion every time such an effect is needed, we can instead re-use a standard part, originally designed for a particular context. Indeed, the fact of re-use of a single design, instead of independent origination of functionally equivalent designs, means we often see idiosyncratic features preserved in quite different artefacts; for example, the same manufacturer's name appears on the electrical motors inside quite different tools and gadgets. If natural selection is responsible for organic adaptation, then although similar functional requirements might bring broadly similar structures into existence in distantly related species, we should still expect significant differences in incidental features of these similar structures, just as independently invented motors bear different makers' names. But if intelligent design is responsible for organic adaptation, then we should expect to find that when functional requirements are similar, structures will be similar even in idiosyncratic ways. On this view, the intelligent designer is like an animal breeder adept in the use of genetic engineering, one who can transport valuable traits directly from one species to another, idiosyncrasies and all, even when the two species are wholly unrelated. Intelligent design appears to predict that all wings, for example, will be built to a standard design, regardless of whether we are looking at dinosaurs, mammals or birds.

It turns out that the wings of birds, bats and pterosaurs have considerable structural differences, even though they are all built for flight. The same goes for the eyes of mammals, squid and insects. Of course the intelligent design theorist can respond in ways that make intelligent design compatible with these observations: perhaps the designer, like us, gets bored with using the same design for the same purpose; perhaps the designer is gifted enough to build different structures to reflect differences in functional requirements in different species. Perhaps several designers are at work, with different ones assigned to different groups of species. But securing compatibility between intelligent design and the observations we make is not the same as generating successful predictions from intelligent design; rather, it is denying that intelligent design predicts what one might think it predicts, while refusing to comment on what it does predict.

If the preceding arguments are right, then the main problem with the intelligent design hypothesis is simply that there is so little evidence in its favour. It makes no successful predictions, it fails to unify diverse classes of phenomena, and it has garnered no support for the alleged character and abilities of the designing agent or agents. It is on a par with the hypothesis of disco-dancing fairies, invoked to explain the drumming noise coming from my roof. This leads me to question Dembski's comments on the scientific spirit:

> Science is supposed to give the full range of possible explanations a fair chance to succeed. That's not to say that anything goes; but it is to say that anything might go. In particular, science may not, by a priori fiat, rule out logical possibilities. Evolutionary biology, by limiting itself to exclusively material mechanisms, has settled in advance the question of which biological explanations are true, apart from any consideration of empirical evidence.
>
> (Dembski 2004: 329)

The fairy hypothesis makes the drumming noise likely, but suppose we refuse to take it seriously in spite of this. Is this to rule it out 'by a priori fiat'? No: our dismissal is not wholly a priori, because it is justified by the fact that we have no evidence in favour of it, even though dancing fairies provide a possible explanation for our observations. Science's opposition to fairies is an opposition based on empirical data. Similarly, science rejects intelligent design because the hypothesis that intelligent design is responsible for the structure of the flagellum has negligible evidential support. Science should not be as even-handed as Dembski seems to suggest.

9. DARWIN AND RELIGION

The hypothesis that organic adaptation is produced by a designing intelligence is without merit. Intelligent design is a hopeless theory. That is why it should not be taught in school biology classrooms: let it in, and we will also have to require school geography teachers to cover the dancing fairy hypothesis as well as standard meteorological work on rain.

In exposing the shortcomings of intelligent design one does not thereby demonstrate that Darwinians should be atheists. Consider the analogous case of Newton's laws. Why do objects fall to the ground? On the one hand, we might cite gravity. As an alternative, we might cite the will of an intelligent creator. It would be absurd if school physics classes had to give these two theories even-handed treatment. Even so, we might hold that there is a form of intelligent design that explains Newton's laws themselves. Perhaps an intelligent designer is responsible for the fact that the universe follows regularities of the sort that Newton exposed. This was, for a time at least, Darwin's own conception of God: not a vulgar plate-spinner, ceaselessly intervening in worldly affairs to produce a species here, or fine-tune an adaptation there, but a God who sets up a small number of elegant laws, which by their own action produce the full range of phenomena we see around us. Intelligent design's image of a creator who fiddles with bits of bacteria is, Darwin writes, 'beneath the dignity of him, who is supposed to have said let there be light & there was light.—' (*Notebook* D, quoted in Barrett *et al.* 1987).

Darwin rightly rejected the argument from individual biological adaptations to a divine designer. He was, for a time, more taken with the argument from regular and fecund laws to a designing intelligence. This argument has its own problems, similar to the ones we have already encountered in the argument from adaptation to intelligence. What reasons do we have to countenance the existence of an intelligent agent, beyond the fact that such a thing would explain the lawfulness of the universe? Why think laws of nature stand in need of explanation at all? Is it because without a designer a lawful universe would be highly unlikely? Does it make sense to talk about probabilities here? My own view is that problems like these undermine the argument from laws to design. But these questions are as old as the hills. They do not concern the specific relationship between Darwin and religion.

SUMMARY

Darwin's argument in the Origin is structured in such a way that it conforms to John Herschel's vera causa standard for a respectable scientific theory. This standard recognises the shaky support conferred on a theory when it does nothing more than make sense of some limited set of observations. False theories, as well as true ones, can have a good explanatory fit with a limited body of data. This means that in addition to this explanatory fit between theory and data, further conditions need to be met before we have strong evidence in favour of the theory. Herschel himself suggests several plausible conditions that might meet this challenge: perhaps the theory needs to make sense of diverse phenomena; perhaps we need 'direct' evidence of some kind in favour of the suppositions of the theory itself; perhaps we need 'direct perception' of the causes posited by the theory. Darwin attempts to meet all of these standards. He shows how his theory is able to explain phenomena in the domains of embryology, classification, the distribution of species around the globe, the anatomy of different species, and so forth. He argues that we have direct evidence in favour of the existence of the variation and inheritance that natural selection demands. And he argues that we have direct experience of a force analogous to natural selection, in the shape of artificial selection by human breeders. Darwin is well aware of the hurdles an explanatory theory must clear if we are to take it seriously. Like Darwin, modern intelligent design theorists also argue that we should believe a hypothesis in virtue of its explanatory power. But modern intelligent design theory, unlike Darwin's theory, fails miserably on all plausible accounts of what kinds of additional evidence are required before we take an explanatory theory seriously. This does not show that there is no room for reconciliation between Darwinism and religion. It does, however, show that intelligent design is not a credible scientific theory.

FURTHER READING

Readers will get a good sense of the general force of Darwin's case in favour of evolution by reading the final chapter (fourteen) of the *Origin*.

Useful papers on Darwin, Whewell and Herschel include:

Ruse, M. (1975) 'Darwin's Debt to Philosophy: An Examination of the Influence of the Philosophical Ideas of John F. W. Herschel and William Whewell on the Development of Charles Darwin's Theory of Evolution', *Studies in History and Philosophy of Science*, 6: 159–81.

Hodge, M. J. S. (1977) 'The Structure and Strategy of Darwin's "Long Argument"', *British Journal for the History of Science*, 10: 237–46.

Waters, K. (2003) 'The Arguments in the Origin of Species', in J. Hodge and G. Radick (eds) *The Cambridge Companion to Darwin*, Cambridge: Cambridge University Press.

Darwin's theory of pangenesis is introduced and explained in:

Endersby, J. (2003) 'Darwin on Generation, Pangenesis and Sexual Selection', in J. Hodge and G. Radick (eds) *The Cambridge Companion to Darwin*, Cambridge: Cambridge University Press.

For a general treatment of *Inference to the Best Explanation* see:

Lipton, P. (2004) Inference to the Best Explanation, second edition, London: Routledge.

My assessment of the modern design argument is heavily influenced by Elliott Sober. A useful overview of his position on these matters can be found in a recent collection of articles on design (and readers will get a good sense of the views of Dembski and Behe by looking at their contributions to this volume also):

Sober, E. (2004) 'The Design Argument', in W. Dembski and M. Ruse (eds) *Debating Design: From Darwin to DNA*, Cambridge: Cambridge University Press.

On Darwin and religion in general, readers might look to:

Brooke, J. (2003) 'Darwin and Victorian Christianity', in J. Hodge and G. Radick (eds) *The Cambridge Companion to Darwin*, Cambridge: Cambridge University Press.

Ruse, M. (2000b) *Can a Darwinian be a Christian? The Relationship between Science and Religion*, Cambridge: Cambridge University Press.

One topic that is not covered in this book is the evolution of religion itself. Evolutionary explanations for the origin and prevalence of religious views can be found in:

Wilson, D. S. (2002) *Darwin's Cathedral: Evolution, Religion and the Nature of Society*, Chicago: University of Chicago Press.

Dennett, D. C. (2006) *Breaking the Spell: Religion as a Natural Phenomenon*, London: Allen Lane.

Five

Mind

I. SQUANDERED RICHES?

Writing in 1969, the biologist Michael Ghiselin bemoaned the failure of the majority of his contemporaries to take a Darwinian perspective on the mind. At the same time, he expressed concern that the end result of such a shift of attention might be disappointing, especially when compared with Darwin's own achievements in this area:

> It is easy to see how a psychologist, attempting to give evolutionary meaning to his data, would tend to use habits of thought quite different from those employed by Darwin. The natural inclination would be merely to impose an oversimplified evolutionary rationalization upon the observations.
>
> (Ghiselin 1969: 210)

Nearly forty years on, there is no shortage of work that goes on under the banner of 'evolutionary psychology'. Steven Pinker's decision to take 'How the Mind Works' as the title for his commercially successful popularisation of this field shouts the promise that many see in the evolutionary stance (Pinker 1997). But evolutionary psychology has met with stiff opposition in the last fifteen years, much of it recapitulating the debate of the 1970s and 1980s that E. O. Wilson ignited with the publication of his 1975 work Sociobiology. This resistance cannot all be explained away as a struggle over turf. It is true that anthropologists, sociologists and social psychologists often regard evolutionary psychology as embodying an overly simplistic picture of human individuals and human societies. But biologists, too, have opposed evolutionary psychology on occasions, and for just the reasons that Ghiselin foresaw. In some cases, they have not been shy in expressing contempt for the subject. Jerry Coyne, a Professor at the University of Chicago's Department of Ecology and Evolution, has remarked that: 'If evolutionary biology is a soft science, then evolutionary psychology is its flabby underbelly' (Coyne 2000). For Coyne and others, what David Buss (1999) has enthusiastically termed the 'New Science of the Mind' is no science at all.

Out and out opposition to the application of evolutionary knowledge to the minds of humans and other animals seems misplaced. It would be strange if the facts that humans are vertebrates, mammals and primates, that they share common ancestors with these animals, could yield no knowledge about human thought and human feelings. Yet Ghiselin worried in 1969 that psychologists would not be as adept at extracting this knowledge as Darwin was; Coyne is convinced that they have failed in this task. Darwin wrote a lot on human psychology in one of his best-known books, The Descent of Man. We will investigate Descent's work on ethics and politics in considerable detail in chapters six and eight. Here I take the somewhat unusual step of focusing instead on The Expression of the Emotions in Man and Animals. This is justified partly on the grounds of novelty, but primarily because the topic of the emotions gives us particularly rich opportunities for examining Darwin's legacy. The object of this chapter is to lay out the themes of Expression, and to see whether today's evolutionary psychology really measures up so badly to the mark Darwin put down.

2. THE THREE PRINCIPLES OF EMOTIONAL EXPRESSION

Darwin initially planned to include his work on emotional expression as part of The Descent of Man, but decided in the end to save it for a later publication, which appeared in 1872. When Ghiselin evaluated Expression in 1969 it had received comparatively little attention from scientists; that is no longer the case, thanks in large part to work by Paul Ekman, who has devoted his career to a thorough elaboration and defence of views akin to Darwin's, and whose recent edited edition of *Expression* has made it widely accessible (and affordable). Modern Darwinians now routinely praise the work – Richard Dawkins' back-cover blurb on the Ekman edition exemplifies this through some typically partisan scorekeeping: '*Expression* predates Freud, and it will still be illuminating human psychology long after Freud's discrediting is complete'. Yet it is curious how un-Darwinian this book is, at least if we understand Dawkins' own views to define what it means to be a Darwinian today. *Expression* contains barely a single mention of natural selection, and it relies primarily on a Lamarckian mode of inheritance to explain how patterns of expression are transmitted from parent to offspring. Let us begin by exposing the main themes of the book, in order to understand how the likes of Dawkins can praise it so strongly, even though its theories might appear antithetical to modern evolutionary thought.

Darwin uses three principles to explain the expression of emotions. They are the *Principle of Serviceable Associated Habits*, the *Principle of Antithesis* and the *Principle of Direct Action*. The basic gist of the first principle is simple to grasp. Some emotional expressions begin as functional responses to external stimuli, and they continue to be inherited and manifested in response to characteristic stimuli, whether they remain functional or not. Darwin uses this principle to explain the expression of fear in humans. Initially, he says, some early animal ancestor of humans, when confronted with an enemy, may have willed its hair or feathers to puff up, thereby making itself appear larger and more intimidating. The Lamarckian element to Darwin's theory can be found in the appeal he then makes to 'use-inheritance'. Darwin believed that a creature that initially willed some action would eventually produce the same behaviour habitually and automatically. Finally, the behaviour would be passed on to the creature's offspring as instinct. A population of birds that tried to puff their feathers up to make themselves look bigger would eventually produce young that automatically responded to enemies in this way. Darwin regards the fact that human hairs stand on end when we are fearful as the product of this kind of inheritance, even though he believes such a response no longer has a useful function for us, our hairs having no perceptible effect on visible size. Darwin adds that as time goes by the initially functional fear response (puffed up hairs or feathers) is triggered not only by an enemy, but by the mere thought of danger.

Darwin thinks the expression of fear in humans is functionally neutral. He believes that some emotional expressions – such as surprise – remain functional in humans. When we are surprised our eyes open wide, allowing us to see more easily the unexpected object or event (Expression: 280–81). In other cases Darwin argues that our emotional expressions are now detrimental to survival. In trying to understand the behaviour of someone who is mortally terrified, Darwin begins by noting that running from a fearsome enemy, or violent struggle with the enemy, is a common functional reaction. He continues:

> As these exertions have often been prolonged to the last extremity, the final result will have been utter prostration, pallor, perspiration, trembling of all the muscles, or their complete relaxation. And now, whenever the emotion of fear is strongly felt, though it may not lead to any exertion, the same results tend to reappear, through the force of inheritance and association.
>
> (Ibid.: 308–9)

That is why we may fall to the ground and tremble when terrified, even though this may be the worst thing to do in the face of great danger.

It is worth saying a little in defence of Herbert Spencer at this point. Spencer is sometimes presented as the enemy of wisdom, with Darwin as its champion. What is sensible in evolutionary biology we owe primarily to Darwin; what is foolish we owe to Spencer's distortions of Darwin's system. Yet this neglects the influence Spencer appears to have had on Darwin's thought. Near the beginning of Expression Darwin says that Spencer's theory is 'the true theory of a large number of expressions; but the chief interest and difficulty of the subject lies in following out the wonderfully complex results' (ibid.: 16). Darwin goes on to quote from Spencer's own much earlier work (his 1855 *Principles of Psychology*) with approval: 'Fear, when strong, expresses itself in cries, in efforts to hide or escape, in palpitations and tremblings; and these are just the manifestations that would accompany an actual experience of the evil feared . . . ' (ibid.). Any modern reader who is impressed by Darwin's Principle of Serviceable Associated Habits should not be so miserly as to withhold all credit from Spencer; if the reader is unimpressed, then Spencer should not shoulder all the blame.

Darwin's second principle, the *Principle of Antithesis*, is harder to understand than the first. He applies it when discussing affection in dogs. Darwin begins by outlining the aspect of a dog in a 'savage or hostile frame of mind.' The dog:

> walks upright and very stiffly; his head is slightly raised, or not much lowered; the tail is held erect and quite rigid; the hairs bristle, especially along the neck and back; the pricked ears are directed forwards, and the eyes have a fixed stare.
>
> (Ibid.: 55–56)

These actions, Darwin thinks, are explicable through the first principle, for they can be understood as functional accompaniments of an intention to attack. But imagine the dog in question realises that the apparent enemy:

> . . . is not a stranger, but his master; and let it be observed how completely and instantaneously his whole bearing is reversed. Instead of walking upright, the body sinks downwards or even crouches, and is thrown into flexuous movements; his tail, instead of being held stiff and upright, is lowered and wagged from side to side; his hair instantly becomes smooth; his ears are depressed and drawn backwards, but not closely to the head; and his lips hang loosely.
>
> (Ibid.: 56)

Darwin cannot see how these expressions might be functional accompaniments to affection, and proposes instead that they are associated with affection only because they are the opposite (the antithesis) of the expressions associated with affection's own opposite emotion.

Darwin recognises that his second principle remains incomplete. Why are animals set up in such a way that affection results in a suite of expressive muscular responses that oppose those of hostility? What explains the association between opposite emotions and opposite reactions? Here Darwin draws on humans' and animals' daily experience of the physical world. He begins by noting that we are accustomed to bringing about opposite effects on physical objects by using our muscles in opposite ways. We need to push to make something go away from us, and pull to make it come towards us. Over time, Darwin believes this practice of reversing our muscular efforts when our intentions are reversed becomes habitual and instinctive. He claims the same is true for animals. This is why, when a dog wishes to show affection, it automatically produces a suite of responses that are the muscular opposites of hostility (ibid.: 66–67).

On the way to formulating this explanation for the Principle of Antithesis, Darwin considers another. He notes that:

> As the power of intercommunication is certainly of high service to many animals, there is no *a priori* improbability in the supposition that gestures manifestly of an opposite nature to those by which certain feelings are already expressed, should at first have been voluntarily employed under the influence of an opposite state of feeling
>
> (Ibid.: 63)

If a dog is to signal to its master that it is affectionate, then that signal can be achieved with the least ambiguity if it is entirely opposite to the expression of hostility. Darwin rejects this explanation because he thinks it demands an implausible level of conscious awareness on the part of the dog. He thinks the dog would need not only an intention to show its master that it is friendly, but an intention to signal this using a suite of behaviours chosen *because* they are in opposition to the behaviours associated with hostility. Although Darwin is willing to credit dogs with intentions of the first kind (and such an intention is involved in the explanation of the Principle of Antithesis that he finally endorses), he does not believe that a dog is capable of forming a complex communicative intention of the second type (ibid.: 66).

Some of Darwin's modern-day champions have been too keen to cast him as a pioneer of our own best ideas. Suzanne Chevalier-Skolnikoff (1973: 20) writes that: 'All the recent investigators of facial expressions either implicitly or explicitly agree with Darwin that the functions of these expressions are communicative, and that such communications regulate social behavior'. This recent work is certainly of considerable importance, but as we have seen Darwin was sceptical of the communicative function. He rejects it when considering the principle of antithesis, and he only mentions it again very briefly towards the end of the book.

Darwin's third principle – the *Principle of Direct Action* – will not concern us much in this chapter, but it is important to mention, for it demonstrates once again how Darwin does not always seek functional explanations for the behaviours that interest him. This principle is purely mechanical. Darwin accepted what is some-times referred to today as the 'hydraulic' view of the mind. 'Nerve force' was considered a fluid, flowing from the mind, through the nervous system, to the muscles. Any strong excitement could thereby disrupt the flow of nerve force, and give rise to muscular activity. Nothing more was required to explain why the most energetic of emotions – great fear, but also great anger and joy – should cause trembling in the sufferer. Just as one should not ask what the function is of vibration in an engine, so one should not look to give a functional explanation for trembling in the fearful. A greatly excited mind will inevitably produce a degree of juddering in the organism as a whole.

Why does Darwin make only intermittent reference to natural selection when laying out his three principles of emotional expression? He is not opposed in general to explaining mental traits by reference to selection. He argues in *Descent* and the *Origin* that natural selection is capable of explaining instincts. Darwin is explicit in *Expression* that some emotional expressions were, at one time, useful. Presumably, then, they were beneficial in the struggle for existence. And he agrees, of course, that offspring inherit the emotional expressions of their parents. Are not these conditions sufficient for us to say that natural selection explains emotional expression? The fact that for Darwin they are not shows that he does not understand natural selection in the way that modern biologists tend to. None of *Expression*'s three principles explains the emergence of emotional responses by appealing in an essential way to the accumulation of *successive, slight* variations. The Principle of Direct Action explains trembling as a necessary mechanical by-product of an excited mind. The Principle of Serviceable Associated Habits appeals to such things as a bird's intentional puffing up of its feathers to explain why this reaction is associated with fear. The Principle of Antithesis appeals to a dog's habitual tendency to reverse the action of its muscles when its intentions are reversed to explain its expression of affection.

Gradual variation, and not merely the 'heritable variation in fitness' we met in chapter two, is, for Darwin, the key to the propriety of an appeal to selection. In *Descent*, Darwin describes a handful of behaviours that become instinctive owing to their initial conscious performance, which is followed by their inheritance as they become habitual. He then invites us to agree with him that 'the greater number of the more complex instinct appear to have been gained in a wholly different manner, through the natural selection of variations of simpler instinctive actions' (*Descent*: 88). This is why he makes room for an appeal to selection when discussing the Principle of Associated Serviceable Habits:

> It further deserves notice that reflex actions are in all probability liable to slight variations, as are all corporeal structures and instinct; and any variations which were beneficial and of sufficient importance would tend to be preserved and inherited . . . [A]lthough some instincts have been developed simply through long-continued use and inherited habit, other highly complex ones

have been developed through the preservation of variations of pre-existing instincts—that is, through natural selection.

(*Expression*: 47)

Later on he points out in a similar vein that selection may explain the puffing up of hairs or feathers associated with fear and anger, although once again this is only a passing note at the end of a long section in which Darwin is concerned primarily with building a case for how animals might consciously will such a fearsome countenance:

> Nor must we overlook the part which variation and Natural Selection may have played; for the males which succeeded in making themselves appear the most terrible to their rivals, or to their other enemies, if not of overwhelming power, will on an average have left more offspring to inherit their characteristic qualities, whatever these may be and however first acquired, than have other males.

(Ibid.: 107)

Darwin is reminding us here that even if he is wrong in his speculation that our ancestors may have willed their feathers or hair to stand erect when confronted by enemies, natural selection may be able to explain this expression of fear instead.

3. COMMON ANCESTRY

If Darwin's basic ideas seem so different from those of many modern Darwinians, one might wonder how his work has been able to impress them so much. The answer can be best appreciated when we see two of Darwin's primary objectives in writing *Expression*. First, he is keen to mount a case against Charles Bell (probably best known these days as author of *The Hand*, another of the *Bridgewater Treatises*), whose *Anatomy and Philosophy of Expression* defended the view that humans have facial muscles specially created for the purpose of emotional expression. Darwin argues, contrary to this, that humans in fact share expressive muscles, and patterns of muscular expression, with animals. The attention to detail in this work, which ranges over several species, remains a model of comparative anatomy (Chevalier-Skolnikoff 1973).

The shared musculature that Darwin points out does not, by itself, undermine Bell's position. Similar muscles, all suited to emotional expression, might have been specially built for the purpose in several species, including humans. But this cannot explain the apparent arbitrariness in human emotional expression. Enraged dogs fight with their teeth. Enraged humans do so very rarely, but they show their teeth all the same (*Expression*: 240). Why would humans have muscles specially created for such a useless role? We can most easily understand our tendency to show our teeth when angry as an inheritance from a much earlier ancestor (common to humans and to dogs) which did use its teeth for fighting. Darwin's argument and his evidence in favour of our having inherited our emotional expressions from some distant ancestor can move those who disagree with his more Lamarckian claims about the specifics of how expressions first arise, and how they are inherited.

A second objective for Darwin's book as a whole was to establish the close biological relationship between the different human races. Darwin thought that sexual selection had led the human races to diverge from each other in various ways, but he believed that humans were all members of one single species, and that the common ancestors of all races were similar to modern humans. He attempted to show this by demonstrating similarities in emotional expression of humans from all over the world, thereby providing:

> . . . a new argument in favour of the several races being descended from a single parent-stock, which must have been almost completely human in structure, and to a large extent in mind, before the period at which the races diverged from each other.
>
> (Ibid.: 355)

Darwin tries to establish that, regardless of culture, humans express themselves in many similar respects. His evidence was obtained by sending questions to an international army of correspondents, whose responses increased his confidence in the universality of many expressions:

> These statements, relating to Europeans, Hindoos, the hilltribes of India, Malays, Micronesians, Abyssinians, Arabs, Negroes, Indians of North America, and apparently to the Australians— many of these natives having had scarcely any intercourse with Europeans—are sufficient to show that shrugging the shoulders, accompanied in some cases by the other proper movements, is a gesture natural to mankind.
>
> (Ibid.: 269)

This evidence from cultures that had had little contact with Europeans is important, for it counts against the hypothesis that human cultures owe their similarities in emotional expression to recent learning from each other, rather than to more ancient inheritance from a recognisably human ancestor. Darwin mentions natural selection for a final time in *Expression* to dismiss it as an explanation for the trans-cultural resemblance he has tried to establish:

> No doubt similar structures, adapted for the same purpose, have often been independently acquired through variation and Natural Selection by distinct species; but this view will not explain close similarity between distinct species in a multitude of unimportant details.
>
> (Ibid.: 355)

Once again, Darwin's evidence for the universality of emotions remains suggestive to modern readers, and his inference to common ancestry as the explanation for this universality is not dependent on the theory of inheritance to which one is committed.

4. THE UNIVERSALITY OF EMOTIONAL EXPRESSION

I wrote that today's psychologists have viewed Darwin's case for the universality of emotional expression as suggestive; they have not, however, viewed Darwin's evidence as watertight. Paul Ekman has tried to bolster Darwin's claim with additional empirical work, and it is perhaps here that *Expression* has been most fertile (Ekman 1973). Darwin, as I noted, set as one of his tasks in *Expression*:

> [T]o ascertain whether the same expressions and gestures prevail, as has often been asserted without much evidence, with all the races of mankind, especially those who have associated but little with Europeans. Whenever the same movements of the features or body express the same emotions in several distinct races of man, we may infer, with much probability, that such expressions are true ones—that is, are innate or instinctive. Conventional expressions or gestures, acquired by the individual during early life, would probably have differed in the races, in the same manner as do their languages.
>
> (*Expression*: 22)

On the hypothesis that emotional expressions are learned conventions, we would expect them to differ from culture to culture, in just the same way that the convention for which word is used to refer to trees differs across cultures ('tree' in English, 'Baum' in German, 'arbre' in French). So if we discover that diverse cultures express emotions in the same ways, this is evidence against the hypothesis that they are conventions learned during infancy.

One problem that Ekman (1973) diagnoses in Darwin's method of gathering data is his use of leading questions. At the beginning of *Expression* Darwin gives us a numbered list of questions he dispatched to correspondents around the globe. Examples include:

> (1) Is astonishment expressed by the eyes and mouth being opened wide, and by the eyebrows being raised? . . .

> (3) When a man is indignant or defiant does he frown, hold his body and head erect, square his shoulders and clench his fists?
>
> (*Expression*: 22)

It is clear enough what answer Darwin is expecting from these questions ('Yes'), and the responses that came back from Darwin's informants ('several of them missionaries, or protectors of the aborigines' [ibid.: 24]) may have been efforts to please their famous correspondent, or reflections of their own European preconceptions of emotional expression projected onto the people they lived with.

Ekman's research was initially limited to students from Japan, Brazil, China, Argentina and the United States. He gave the students a series of pictures, showing what Ekman had determined to be expressions characteristic of six different emotions (happiness, sadness, anger, fear, surprise and disgust). The students were also given words for different emotions in their own languages, and they were asked to match the words to the pictures. Ekman found that Japanese students, for example, would pair the Japanese word that best translates 'fear' to the image that American students also picked out as showing fear, and so on for each of the six emotions he tested, and for each of the groups of students.

By itself, this experiment does not rule out the possibility that the same emotions are associated with the same expressions because of shared learning experiences, to which all the students had been exposed. Perhaps global film and television coverage enable Jack Nicholson's look of happiness and Paul Gascoigne's look of sadness, to be learned the world over. Ekman thus turned his attention, just as Darwin had done, to cultures that have had little contact with Westerners. He looked to the Fore of New Guinea. The problem for his earlier experimental method was that the Fore has no written language. Ekman instead produced photographs of three different facial expressions, which he gave to the Fore research subjects. A story would be told (in the Fore language). Here is an example of one of the stories:

> She is sitting in her house all alone and there is no one else in the village; and there is no knife, ax, or bow and arrow in the house. A wild pig is standing in the door of the house and the woman is looking at the pig and is very afraid of it. The pig has

been standing in the doorway for a few minutes and the person is
looking very afraid and the pig won't move away from the door
and she is afraid the pig will bite her.

(Ekman 1973: 211)

After the story was read, the subject would be asked which of the
pictures shows the expression of the woman in the story. In
general, the Fore people tended to pick the same image that
Ekman's American subjects picked when they were read the story
in English. Ekman concludes, on the basis of this and other experiments, that Darwin was broadly right. In summary, 'the same
facial expressions are associated with the same emotions, regardless of culture or language' (ibid.: 219).

5. CULTURE AND THE EVOLUTIONARY APPROACH

Does the evolutionary approach to the emotions, exemplified by
Darwin and Ekman, ignore the importance of cultural variation
and cultural influence? Does it downplay them in some sense, or
perhaps show culture to be less important than we might otherwise have thought?

The first thing to note is that neither Darwin, nor Ekman,
claims that emotional expressions are universal in the sense that
every person in the world expresses emotion in the same way. To
state the obvious, in all cultures there are individuals with very
unusual facial musculature, or very limited control over their
facial muscles, who do not express emotions in the same ways as
the majority. Darwin and Ekman claim, instead, that emotional
expression is *pan-cultural*: the same patterns of emotional expression
can be found fairly reliably in every culture.

The second point worth noting is a difference between
Darwin's account of emotional expression and Ekman's. Darwin
subscribes, implicitly at least, to a fairly strong distinction
between emotions themselves (fear, joy), and the outward signs
of emotions (trembling, smiling). Ekman thinks of emotions as
what he calls 'affect programs'; rather than drawing a distinction
between fear, say, and its mode of expression, Ekman thinks fear

simply is the suite of responses, including facial responses, triggered by stimuli that are perceived to be dangerous.

Darwin's arguments against Bell, and in favour of his view about the relatedness of human races, require only that certain patterns of facial muscular activity are common to diverse human cultures, and to non-human species. This does not entail that emotions themselves are universal. But Ekman believes that fear is nothing more than a suite of responses, and he has shown that particular facial expressions are universally associated with particular kinds of stimuli. So Ekman, unlike Darwin, explicitly claims that emotions themselves (or at least what he recognises as the six 'basic emotions' of happiness, sadness, anger, fear, surprise and disgust) are universal.

Does Ekman's view of the universality of emotions (where these are understood as affect programs) deny the facts of cultural variation? It does not, and for a variety of reasons. First, he does not claim that all cultures react in the same ways to the same stimuli. True enough, he says that they will tend to react in the same ways to stimuli perceived as dangerous, but that is compatible with different cultures holding quite different things to be dangerous.

We can think of an 'affect program' as a routine that takes stimuli characterised in a certain way (e.g. dangerous things, in the case of fear), and which yields characteristic expressions as outputs. Ekman acknowledges, as we just saw, that culturally specific factors may determine what sorts of phenomena are fed into each program – what sorts of events, for example, are characterised as dangerous. Ekman also acknowledges that there can be culturally specific determinants of the outputs of affect programs, which he calls 'display rules, norms regarding the expected management of facial appearance' (Ekman 1973: 176).

To see what sorts of things these display rules are, consider this apparent counter-example to the universality of emotional expression: when Samurai women learn that their husbands or sons have died in battle, they show expressions that Europeans would regard as signs of joy. In fact, this is no counter-example at all, for as Ekman notes, this is not a culture in which smiling is an expression of grief; rather, it is a culture like ours in which smiling is a sign of joy. But unlike our culture, there is a local expectation that grief should be suppressed, and joy signalled instead. (There is another explanation: namely, that Samurai women are genuinely joyful when their loved ones die in battle. If that is the case then once again it is no counter-example, for it is not a case where smiling is the sign of grief.)

One might worry that the invocation of display rules is a plainly *ad hoc* manoeuvre, which protects the theory against all potentially problematic data. Such worries are misplaced. In another experiment, Ekman was able to manipulate these cultural display rules; he did so by showing a film with repugnant scenes to a group of Japanese students, and then to a group of American students. When the students were left alone in the viewing room, and recorded secretly, Ekman found that both groups showed the typical facial expressions associated with disgust. Groups then viewed the film with an experimenter from their own culture, and were asked to describe their feelings as they watched the film. This time, while the Americans continued to show a strong disgust reaction, the expressions of the Japanese were far more positive. Later, when Ekman was able to use a slow motion camera, he found that even in the presence of the experimenter the Japanese students began to form expressions characteristic of disgust, which were then followed by more neutral expressions. Such data are just what the existence of a universal disgust affect program, coupled to a culturally specific display rule, would predict.

How can we reconcile Ekman's work, which apparently establishes the universality of affect programs, with work done by anthropologists that asserts the cultural specificity of many emotions? A good example to focus our discussion, one which I have borrowed from a useful article by Mallon and Stich (2000), comes from anthropologist Catherine Lutz's work on the Ifaluk people from the Caroline Islands in Micronesia. According to Lutz, the Ifaluk recognise an emotion that they call 'song' (Lutz 1988). Song is like anger in some ways, but unlike anger song must come from a morally justified cause. Whether Fred is angry is entirely decided by facts about Fred. Yet Fred cannot be *song* unless he has been genuinely wronged. *Song* is a little like knowledge in this respect. You might be thoroughly convinced you know that Tallinn is the capital of Latvia, but unless Tallinn is the capital of Latvia, you do not, in fact, know this. Similarly, you could be convinced you are *song*, but unless your ire is morally justified, you are not *song*, no matter how worked up you might be.

One might be tempted to reason like this. There is a deep conflict between anthropological claims about cultural specificity, and evolutionary claims about cultural universality. For Lutz's picture suggests that *song* does not exist in European cultures, and anger does not exist in the Ifaluk culture. Ekman, meanwhile, appears committed to the view that anger exists in all cultures, Ifaluk included.

In fact, as Stich and Mallon argue, we can accept a large amount of what both Ekman and Lutz say. Ekman claims that a small number of affect programs are universal. This entails that the Ifaluk have the anger affect program. But it does not entail that they have any concept of the anger affect program, and it does not entail that recognition of the anger affect program as such plays any role in their social interactions. Ekman's view can be made compatible with that of Lutz if we hypothesise that the Ifaluk use concepts that have their proper application only to affect programs when they are triggered in particular ways. On this view, *song* is something like the-anger-affect-program-when-triggered-by-a-justified-cause.

There is no conflict between Ekman's claim that affect programs are universal, and the claims of anthropologists like Lutz that specific cultures have their own peculiar emotion concepts which classify states of mind in ways quite alien to those of Europeans, whose use plays roles of unique local significance, and which strain to be translated into European languages in virtue of these features. Ekman's description of some of his experiments does threaten to bring him into genuine conflict with Lutz regarding the universality of emotion concepts: 'In every culture we studied, the observers were given the words for these emotions [fear, anger, happiness etc.] in their own language and were required to choose one word for each picture' (Ekman 1973: 198). This appears to commit him, unwisely, to the claim that every culture has a concept of anger, for how else could the experimenter give people from all the cultures he studied 'words for' anger or happiness 'in their own language'? But Ekman is not required to defend any claim about the universality of the *concept* of anger if his work on the universal existence of the anger affect program is to stand. (Analogously, a cognitive psychologist might argue that short-term memory is a universally present element of the human mind, without claiming that all cultures have the concept of short-term memory.)

Sometimes hostility to evolutionary psychology comes from a suspicion that what we can establish as universal across human cultures will need to be so thinly characterised as to offer scant explanatory resources to those interested in studying human populations. The anthropologist Clifford Geertz states this view forcefully:

> There is a logical conflict between asserting that, say, 'religion', 'marriage', or 'property' are empirical universals and giving them very much in the way of specific content, for to say that they are empirical universals is to say that they have the same content, and to say they have the same content is to fly in the face of the undeniable fact that they do not.
>
> (Geertz 1973: 39–40)

If Ekman has indeed established the existence of culturally universal affect programs – and the evidence that he has done so is strong – then one should not overstate how thin the characterisation of universal traits must be. But perhaps we should interpret Geertz's complaint in a different way: the real worry is not that *psychological traits* are universal only if characterised in a thin way, but that the *concepts* used by humans are universal only if characterised in a thin way. Assuming Ekman is right, the British and the Ifaluk all have the anger affect program. But the only way one could claim that the British concept of anger is the same as the Ifaluk concept of *song* would be by ignoring a huge number of *song*'s most important features.

Anthropologists are often interested in showing how different concepts possessed by different cultures will lead to very different patterns of behaviour and interaction in those cultures. This makes them frustrated with evolutionary psychologists' claims regarding the universality of various traits. 'So what,' an anthropologist might say, 'if the Ifaluk have the anger affect program? This is no big deal, because the Ifaluk themselves do not recognise this affect program. If we are to understand the lives of the Ifaluk, we need to understand, for example, how they use the culturally-specific concept of *song*. We need to ask what kinds of causes are viewed as morally justifying; hence what causes can make one *song*. We need to ask how someone who is *song* will be permitted to treat others (including the person towards whom *song* is directed). We need to examine the conduct of debates over who is *song* and who is not, and the penalties for acting in a *song* manner when one is not *song*. Understanding the cultural specifics of *song* can thereby afford an insight into Ifaluk culture. Even if the facial expression of one who is *song* is illuminated by Ekman's work, the broader politics of *song* in Ifaluk society will be left largely untouched by it.'

How should we respond to this frustration? The question of how much value there is in the evolutionary perspective will depend, at least in part, on which questions we are asking. In the case of *song*, many of the questions a social anthropologist would be likely to ask about its role in Ifaluk society will probably not be illuminated by evolutionary study. But this does not trivialise Ekman's results – it is neither obvious, nor does it appear to be false, that humans in all cultures share affect programs that can be reasonably richly characterised. The fact that anthropologists are not much interested in this result does not make the evolutionary perspective worthless to everyone. That perspective has implications regarding the question of whether patterns of emotional expression are learned conventions, and it suggests that these expressions originated in some common ancestor of all human cultures.

6. THE SANTA BARBARA SCHOOL

So far, I have concentrated on Darwin's own psychological work, and then on the recent work on emotions that *Expression* has inspired. Ekman's is only one way of taking an evolutionary view of the mind, and it is certainly not the best-known. Far more widely discussed in philosophical and popular circles is the methodological stance developed primarily by John Tooby and Leda Cosmides (1992), popularised by Steven Pinker (1997), and brought to students in textbook form by David Buss (1999). When many people refer to 'evolutionary psychology', it is work following the tradition laid down by Cosmides and Tooby which they have in mind. I will refer to this tradition as the *Santa Barbara School*, after Cosmides and Tooby's employer, the University of California at Santa Barbara. This group is committed to several claims, but the following three are the most salient for our purposes (see also Gray *et al*. 2003):

A single human nature: all non-pathological human minds have the same collection of adaptations, fashioned by natural selection.

The adaptive heuristic: reflection on the past demands of the environment in which humans evolved helps us to understand how our minds work today.

Massive modularity: the human mind is like a Swiss-army knife. It is not a general-purpose thinking machine; rather, it is composed of many distinct tools, or 'modules', each adapted to a particular kind of cognitive problem. These modules are innate.

Claims about the reality, universality and character of human nature sometimes play important roles in ethics and political philosophy. The Santa Barbara School's view about the unity of human nature is therefore of considerable philosophical interest, for perhaps it will give an empirical boost to the claims of one or another philosophical camp. The remainder of this chapter examines the unity of human nature and the adaptive heuristic. Discussion of the alleged differences between male and female minds will be reserved for chapter eight, on politics. Discussion of massive modularity is saved for chapter seven, on knowledge.

7. A SINGLE HUMAN NATURE?

The Santa Barbara School believes that human minds are composed of adaptations, fashioned by natural selection. But if we accept this, should we also agree with its claim that all human minds contain the same adaptations?

We saw in chapter three that natural selection maintains adaptive polymorphisms in many species. Several fairly distinct forms can co-exist, often because their fitnesses increase with their rarity, so that when they are unusual their representation in the population is likely to increase, and when they are common their representation is likely to decrease. The result is a mixed population. This was the lesson we learned from the Hawk–Dove model. We also came across the biologist David Sloan Wilson's tentative claim that introversion and extroversion in humans may be alternative adaptations maintained by selection (Wilson 1994). The Santa Barbara School rejects the claim that human minds show adaptive polymorphism. But if adaptive polymorphism is rife throughout the animal and plant kingdoms, why think that the adaptations that underlie human minds must be uniform?

Before examining this issue, it is worth clarifying the Santa Barbara position (and here I am indebted once again to arguments from Buller 2005). It is easy to misunderstand the School's claim that there is a single human nature. The Santa Barbara School is not committed to denying that some people are introverts and others extroverts, for example. They believe that we all share the same cognitive adaptations, but they claim that these adaptations are 'facultative'. Facultative adaptations have a form of flexibility built into them. They are understood to embody developmental 'programs', which specify what sort of mental traits a developing person acquires, depending on what that person's developmental environment happens to be like. To caricature the position for a moment, these programs might embody conditional rules of the form, 'if growing up among aggressive people, become an introvert; if growing up among people easy to dominate, become an extrovert'. Our adult minds are not all the same on this view, but we do share the same cognitive adaptations, understood as developmental programs. When the Santa Barbara School claims there is a single human nature, this is what they mean.

Why does the Santa Barbara School claim that we all share the same facultative adaptations? Could natural selection not have led many different cognitive adaptations (facultative or otherwise) to co-exist in human populations? Cosmides and Tooby think that our knowledge of genetics shows us that this is highly unlikely to have happened. Here is their argument (Tooby and Cosmides 1990): complex adaptations – eyes, wings, cognitive adaptations, too – are built by many different genes acting in concert. If the human population contained individuals with alternative complex adaptations, it would therefore contain individuals with alternative sets of genes required to build those adaptations. But suppose two individuals with alternative complex adaptations were to reproduce; their offspring would have half the genes for one adaptation, and half the genes for the other adaptation. This ill-matched set of genes would not produce any adaptation at all. As they put it, if you mix up half the parts from a Honda, and half the parts from a Toyota, the chances are the car you end up with will not run. So the only way that human development can be made to work is if everyone has genes that specify the same adaptations. That is why human nature must be universal: 'The psychic unity of humankind—that is, a universal and uniform human nature—is necessarily imposed to the extent and along those dimensions that our psychologies are collections of complex adaptations' (Tooby and Cosmides 1992: 79).

At this point we can add another element of complexity to the Santa Barbara position. In principle, they say, facultative adaptations need not take the form 'In environment one, develop psychological trait A; in environment two, develop psychological trait B'. Adaptations can also be tuned to genetic differences, thereby taking the form 'When accompanied by gene one, develop psychological trait A; when accompanied by gene two, develop psychological trait B'. This is why they believe something that might appear contradictory, namely that human nature is universal and there are distinct male and female psychologies. Male and female psychologies, they say, are the alternative developmental outputs of a single developmental program shared across the sexes. These alternative outputs are triggered by the presence or absence of a 'genetic switch' on the Y chromosome.

This helps to answer another query that one might have regarding the Santa Barbara position. Cosmides and Tooby's argument for the improbability of a mixed set of adaptations in the human population does not rely on any specific details of human minds.

If their argument really shows that a complex adaptation can only exist in a population when all of the population's members share the same adaptation, then their argument shows that no species can contain adaptive polymorphism, at least with respect to complex traits. But biologists believe adaptive polymorphism is ubiquitous. So either the majority of biologists are wrong about this, or there is something wrong with Cosmides and Tooby's argument, or the two parties are talking past each other.

It turns out that when Cosmides and Tooby say that human nature is universal, what they have in mind is rather weak, for it is compatible with the existence of genetically distinct types, which have different psychologies because of these genetic differences. They think that males and females are two such genetically distinct types. Their claim that all humans share the same nature in spite of this is largely an artefact of their decision to count alternative genes that send development in one direction or another as external switches directing uniform adaptations, rather than as internal elements of alternative adaptations. So they do not, in fact, deny the possibility of genetically controlled adaptive polymorphism. Even so, they believe that polymorphisms of this kind – the kind that are controlled by genetic switches, rather than environmental cues – are likely to be very rare. This is because a genetic switch commits the developing organism to a particular form (either a male or female anatomy, say) at the beginning of its life. Cosmides and Tooby argue that a 'wait and see' policy, which allows the development of the organism to be directed by the actual demands of the environment, will usually be more effective. But it is highly speculative to base the existence of a universal set of facultative adaptations on a theoretical argument of this kind. After all, consider that humans do not use a 'wait and see' policy in the determination of sex. Sex in our species is determined genetically, according to the presence of X or Y chromosomes. This is not the case for all animals. Several species of turtle, crocodiles and some lizards have no sex chromosomes. Instead, their sex is determined by an environmental switch. More specifically, sex in these species depends on the local temperature during a critical period of embryonic development. Moreover, there are ongoing disputes regarding the adaptive significance of temperature-dependent sex determination, hence it is difficult to pin down any simple set of rules regarding when polymorphism is, and is not, likely to be controlled by genetic switches (see Charnov and Bull 1977; Warner and Shine 2005). These considerations are made more complex in the case of our own species, for it is possible that in some cases humans are able to choose the environments in which they will spend their time, so as to profit from whatever psychology their genes have endowed them with. If this is the case, the advantages of a 'wait and see' policy are undercut (Wilson 1994). It is far from clear, then, that genetic switches should be rare occurrences in the natural world. In short, Cosmides and Tooby's argument falls short of establishing that there is a single human nature in any strong sense.

8. THE ADAPTIVE HEURISTIC

The Santa Barbara School has bold aspirations, namely to reveal the structure and workings of the human psyche – to yield, as Cosmides and Tooby put it, 'The Gray's Anatomy of the Mind'. How can evolutionary thinking help us to do this? The behavioural ecologists John Krebs and Nick Davies introduce the case for how we can expose organic mechanisms in general:

> Visitors from another planet would find it easier to discover how an artificial object, such as a car, works if they first knew what it was for. In the same way, physiologists are better able to analyse the mechanisms underlying behaviour once they appreciate the selective pressures which have influenced its function.
>
> (Krebs and Davies 1997: 15)

Cosmides, Tooby and Barkow seek to apply the same form of thinking to the human mind:

> By understanding the selection pressures that our hominid ances-tors faced—by understanding what kind of adaptive problems they had to solve—one should be able to gain some insight into

the design of the information-processing mechanisms that
evolved to solve these problems.

(Cosmides *et al.* 1992: 9)

In other words, if we want to understand how our minds work
today, then it is important to reflect on the ancestral problems that
shaped them. But what problems are these? According to the Santa
Barbara School, the most salient period for our species' cognitive
evolution is the long stretch of time during the Pleistocene (from
about 1.8 million years ago to 10,000 years ago) when our ancestors lived as hunter-gatherers on the African savannah. Our minds
were shaped to fit the tasks of the hunter-gatherer lifestyle, and
there has not been time enough for evolution to change our
cognitive adaptations since this period ended. The result is that
'our modern skulls house a Stone Age mind' (Cosmides and
Tooby 1997a: 85).

In this section, I want to raise two problems for using the adaptive heuristic. One concerns how much of a leg-up we should
expect reflection on past selective demands to give to the project
of uncovering our current cognitive endowment. The second
concerns whether our minds really are frozen in the Stone Age.

Let us begin with the first problem. In a short article that the
Harvard biologist Stephen Gould wrote in the *New York Review of
Books*, he complained about how little knowledge we have of
ecological conditions in our evolutionary past, and he concluded
that we do not have enough data to predict how our species is
likely to have responded to its past conditions. Cosmides and
Tooby responded forcefully, reminding us that Gould understates
how much we know that may be of relevance:

> Our ancestors nursed, had two sexes, hunted, gathered, chose
> mates, used tools, had colour vision, bled when wounded, were pre-
> dated upon, were subject to viral infections, were incapacitated from
> injuries, had deleterious recessives and so were subject to inbreeding
> depression if they mated with siblings, fought with each other, lived in
> a biotic environment with felids [cats], snakes and plant toxins, etc.
>
> (Cosmides and Tooby 1997b)

How are we to transform this description of our ancestors into a catalogue of the adaptive problems which faced them? Consider the fact that humans lived in an environment with plant toxins. There are several adaptive ways in which humans could react to this. They might acquire gut bacteria that neutralise the toxins; they might avoid plants with high levels of toxins, and eat only those with low levels; they might develop cooking methods that eliminate toxins; and so forth. Which of these is adopted depends not merely on the human environment, but on the range of responses that human psychology and physiology (as well as the nature of existing bacteria) make available to natural selection. So if we are to characterise ancestral environmental problems in a fine-grained way that facilitates novel predictions – not as the problem of plant toxins, but, let us say, as the problem of acquiring gut bacteria for the removal of plant toxins – then we already need data about human nature to use as an input to the adaptive heuristic. These data must, consequently, be generated in a different way, most likely by looking directly at human psychology, physiology, and so forth, or at the psychology and physiology of our close relatives.

This shows that one should not be too ambitious in the value one claims for the adaptive heuristic. The anatomical, physiological, psychological and behavioural features that a species brings to an environment all affect the likely directions the species' evolution will follow. Unless we have fairly rich data regarding a species' anatomy, physiology, psychology and social organisation, we are unlikely to be able to predict its evolutionary response to a past environment. This means that information gleaned from traditional human sciences will retain a leading role in evolutionary attempts to ascertain 'how the mind works' (Sterelny and Griffiths 1999).

What of the second problem? Do we have minds whose organisation has not changed since the Stone Age? I will touch on one issue this raises in chapter seven: if the Stone Age was itself a time of shifting adaptive problems, then perhaps we responded to those problems not by acquiring a set of rigid adaptations, but by acquiring abilities to respond in *ad hoc* ways to new problems as they arose. Maybe we still have malleable minds like this. Since the problems posed by modern living are quite different to those posed in the Stone Age, we should expect a malleable mind to stretch itself in quite different ways now compared with how it was stretched back then. Setting this aside for now, is it true that there has not been time for natural selection to have modified our minds significantly in the last 10,000 years?

Evolution can happen very rapidly. Lake Victoria in Africa contains over 500 different species of cichlid fish. But it appears that around 15,000 years ago Lake Victoria dried up completely. If that is true, it suggests that these 500 species, together with their distinctive adaptations to different niches found in the lake, have all appeared within this short time (Johnson *et al.* 1996). But fish are fish, and they breed much faster than humans. What is more, the interpretation of Johnson *et al.*'s claims is contentious. Even so, comparatively recent changes in human farming practices – specifically the domestication of cattle, and the consequent increase in the use of dairy products – have resulted in a greatly increased incidence of lactose tolerance (a genetically-inherited adaptation) among humans (Richerson and Boyd 2005: 191–92). Here is an example of a shift in the environment since the Stone Age, in the shape of the presence of domestic dairy cows, which has had knock-on effects on our species' genetically controlled adaptations. Of course, the invention of dairy farming is only one of many changes wrought on our environment since the Pleistocene. Many of us now live in cities, we no longer hunt as a matter of necessity, medical technology has improved, and there is no reason to rule out considerable modification of our cognitive adaptations in response to these altered environments, too.

9. DARWIN AND SANTA BARBARA

Darwin's evolutionary psychology is in many ways different to that of the Santa Barbara School. He does not make much use of the adaptive heuristic. He does engage from time to time in what is now called *reverse-engineering*. Here, rather than trying to predict an unknown aspect of current organic makeup on the basis of a past environment, we instead begin with a fairly firm grip on some aspect of current organic makeup and attempt to come to an evolutionary historical understanding of it. As usual, a nice example in *Expression* comes from Darwin's observation of dogs:

> Dogs, when they wish to go to sleep on a carpet or other hard surface, generally turn round and round and scratch the ground with their fore-paws in a senseless manner, as if they intended to trample down the grass and scoop out a hollow, as no doubt their wild parents did when they lived on open grassy plains or in the woods.
>
> (*Expression*: 49)

Darwin explains a great deal about human emotional expression, but he does not do it by predicting what kinds of expressions would have been useful to us in the Stone Age. Instead, he begins by establishing a series of anatomical and behavioural similarities in emotional expression across human cultures and across non-human species (humans and dogs show their teeth when angry). He infers from this that humans and other species owe their expressive similarities to descent from a common ancestor. In biological terms, human fear and dog fear are *homologous*, just like human forelimbs and dog forelimbs. Unlike the wings of birds and bats – which are termed *analogous* – the resemblances between fear in humans and dogs, or forelimbs in humans and dogs, are not to be accounted for by independent evolution. Darwin hypothesises that aspects of these expressions may have arisen as solutions to problems faced in the past (baring of teeth is reverse-engineered as a preparation for attack), and he thereby shows how we can make sense of apparently functionless aspects of human expression.

Darwin's method of illuminating human emotional expression relies on subtle choices regarding which alternative species to investigate. Once a historical hypothesis regarding, say, the origination of anger and its expression in a common ancestor of humans and dogs is established, it should prompt us to ask whether we might not share many other functionally useless, and therefore unsuspected, traits with these relatives. This focus on related species gives Darwin's method a good chance of uncovering unknown facts about the workings of our own minds. Contrast this with recent work by evolutionary psychologists Thornhill and Palmer on rape (Thornhill and Palmer 2000). They argue that the disposition to rape may be an adaptation among human males. They supplement their case with frequent references to their studies of the adaptive advantages of 'forced copulation' in scorpion flies, but there is no suggestion that rape in humans and forced copulation in scorpion flies are traits inherited from a common ancestor with a tendency to either behaviour. Only if rape and forced copulation were homologous should we expect there to be numerous deep similarities between them; the fact that they are not underlines the obvious limitations in applying work on flies to work on the complex human phenomenon of rape.

Darwin's appreciation of the importance of cross-species comparisons is important for a second reason. 'Reverse-engineering' needs to be handled with just as much care as the adaptive heuristic. It is fairly easy to think up lots of alternative past evolutionary scenarios that would explain the traits we see today, including our own cognitive traits (Gould and Lewontin 1979). Darwin's example of the vulture highlights this problem. Vultures eat carrion, and if the vulture's head was covered in feathers the rotting flesh on which it feeds might get stuck in them, encouraging parasites and infection. We might leap to the conclusion that natural selection has designed the vulture's bald head 'for wallowing in putridity' (Origin: 226). Darwin points out, however, that 'we should be very cautious in drawing any such inference, when we see that the skin on the head of the clean-feeding male turkey is likewise naked' (ibid.). This undermines our hasty conclusion, and not merely because it weakens the correlation between baldness and wallowing in putridity. It prompts us to investigate whether vultures and turkeys both owe their bald heads to descent from a common bald-headed ancestor. If they do, and if this ancestor was clean-feeding like the turkey, then we can conclude that selection probably did not modify the vulture's head to enable it to wallow in putridity. This is how comparisons between well-chosen related species can assist us in reverse-engineering. Such comparisons might help us to test the historical assumptions of claims about the circumstances under which emotional expression evolved, too. Darwin's judicious use of cross-species comparisons means that his work remains a model for evolutionary psychology.

SUMMARY

In the past thirty years or so, there has been an explosion of work applying evolutionary theory to the human mind. There is a pedigree for this sort of work in Darwin's own publications: he applied evolutionary ideas to animal behaviours, animal instincts and the human mind. Yet there are significant differences between Darwin's work on the mind, and the mainstream of modern evolutionary psychology. Although Darwin takes an evolutionary perspective on the emotions, natural selection does not come into the foreground in this work. What is more, rather than seeking to explain human emotional expressions in terms of their direct evolutionary functions, he tends instead to explain their features by reference to inheritance from ancestors that we share in common with other species. In spite of these differences, Darwin's work on the emotions has been influential. Many of its main claims have been borne out by subsequent research, especially his claims regarding the universality of emotional expression, which have been bolstered by recent work by Paul Ekman. The best known evolutionary psychological work, however, is not Ekman's, but rather work undertaken by what we here have called the Santa Barbara School of evolutionary psychology. Its members tend to argue that we can best understand how the human mind works by reflecting on how problems posed during the Pleistocene have shaped our species' psychology. They also argue that there is a unitary human nature: a single universal psychology, present in all members of our species. We have seen reasons to be sceptical of both of these claims.

FURTHER READING

The emotions are discussed in great detail in *Expression*, but Darwin also discusses matters of relevance to psychology in chapter seven of *Origin*, and especially in chapters three, four and five of *Descent*. Natural selection features far more prominently in *Origin* and *Descent* than it does in *Expression*.

Robert Richards has written an important historical study of Darwin's view of the mind:

Richards, R. (1987) *Darwin and the Emergence of Evolutionary Theories of Mind and Behavior*, Chicago: University of Chicago Press.

A broad overview of the relevance of Darwinian thinking for contemporary philosophy of mind can be found in:

Sterelny, K. (2003b) 'Darwinian Concepts in the Philosophy of Mind', in J. Hodge and G. Radick (eds) *The Cambridge Companion to Darwin*, Cambridge: Cambridge University Press.

Paul Griffiths's recent philosophical study of the emotions draws on a rich range of sources, and is also a good place to look for more detail on Ekman's work:

Griffiths, P. (1997) *What Emotions Really Are: The Problem of Psychological Categories*, Chicago: University of Chicago Press.

A review of work on the universality of emotion categories can be found in:

Russell, J. (1991) 'Culture and the Categorization of Emotions', *Psychological Bulletin*, 110: 426–50.

On the apparent conflict between evolutionary views and those of social anthropology, Mallon and Stich's article is useful:

Mallon, R. and Stich, S. (2000) 'The Odd Couple: The Compatibility of Social Construction and Evolutionary Psychology', *Philosophy of Science*, 67: 133–54.

There are many works discussing evolutionary psychology, and most are partisan. A balanced introduction to the field as a whole, and the source of the label 'The Santa Barbara School', is:

Laland, K. and Brown, G. (2002) *Sense and Nonsense: Evolutionary Perspectives on Human Behaviour*, Oxford: Oxford University Press.

An important collection of papers that form the foundations of the Santa Barbara School's position is:

Barkow, J., Cosmides, L. and Tooby, J. (1992) *The Adapted Mind: Evolutionary Psychology and the Generation of Culture*, Oxford: Oxford University Press.

Finally, a penetrating assault on the Santa Barbara School is contained in a recent book by the philosopher David Buller:

Buller, D. (2005) *Adapting Minds: Evolutionary Psychology and the Persistent Quest for Human Nature*, Cambridge, MA: MIT Press.

Six

Ethics

1. ETHICS FROM THE SIDE OF NATURAL HISTORY

Darwin remarks at the opening of the fourth chapter of *The Descent of Man* that philosophers of 'consummate ability' have not been short of things to say about our sense of right and wrong. Darwin calls this faculty the 'moral sense', which is 'summed up in that short but imperious word *ought*, so full of high significance' (*Descent*: 120). He sets out to examine the moral sense, partly because it is a human trait of such importance and interest, but also because, as far as he knows, 'no one has approached it exclusively from the side of natural history' (ibid.).

One hundred and four years after *Descent* was first published, the Harvard biologist E. O. Wilson asked his readers to consider something a little stronger, namely, 'the possibility that the time has come for ethics to be removed temporarily from the hands of the philosophers and biologicized' (Wilson 1975: 562). What might it mean to make ethics biological, or to approach ethics from the perspective of natural history?

The philosophers Elliott Sober (1994a) and Philip Kitcher (1993) both outline the different forms such a project might take. The most modest one would expose the evolutionary processes that have led us to make the ethical evaluations we do. A good example might be a natural selection explanation for our tendency to care more about our own offspring than about the offspring of strangers. This is a comparatively modest project because such a historical explanation tells us why we have this tendency, but can remain silent on whether it is a good or a bad tendency for us to have.

The second project is more ambitious, and consists in using evolutionary psychology in a different way. Evolutionary psychology promises, as we saw in the last chapter, to show 'how the mind works'. Consider an ethically fraught issue like overcrowding in prisons. If we think that we ought to reduce prison numbers, one good way of doing this is to reduce crime. If we can understand how the human mind works, then we might also understand the circumstances that lead people to commit crimes, and we could use this knowledge to reduce crime as far as possible. Although it is ambitious, this second project does not seek to use evolutionary thinking to tell us what outcomes we should aim at. Rather, once we decide that we should aim at less crowded prisons, the adaptive heuristic then promises to show how we can best go about doing this. I have already expressed some scepticism about the adaptive heuristic: I will have more to say about this particular way of using it in chapter eight.

For the most part I will be looking at two far more ambitious ways of biologicising ethics in this chapter. I will follow standard terminology and call these 'evolutionary normative ethics' and 'evolutionary meta-ethics'. The distinction between normative ethics and meta-ethics needs some clarification. Normative ethics is probably the subject that most people who have not studied ethics in philosophy classes have in mind when they imagine what the study of ethics must be like. Normative ethics is the study of what we should do, which courses of action are right and wrong, which outcomes are good, and which are bad. These range from quite specific questions to more general ones. Should we allow parents to choose the genetic makeup of their children? Is it better for all members of a society to have equal wealth, equal welfare or equal opportunities? Is an action right if, and only if, it produces the greatest happiness of the greatest number? All of these are questions in normative ethics. Meta-ethics, on the other hand, is the subject that many philosophy students end up spending much of their time on. Meta-ethics is the study of the nature of ethical discourse and its subject matter. Questions in meta-ethics include: Are ethical statements capable of being true or false? If they are sometimes true, what makes them true? Is there a realm of ethical facts? What sort of things might these facts be? Are ethical utterances instead more like expressions of emotion, and if so, what is going on when people disagree about matters of ethics?

Normative ethics and meta-ethics can sometimes have a surprising degree of independence of each other. Suppose, for example, that we do some work in meta-ethics that leads us to conclude that there is indeed some ethical fact of the matter about whether abortion is wrong. This does not settle the normative question of whether abortion is wrong, any more than concluding that there is some fact of the matter about how many blades of grass there are in Hyde Park tells us the number of blades of grass in Hyde Park. Conversely, two people who differ regarding the meta-ethical question of whether there are moral facts might still conduct a fruitful debate over the normative question of whether abortion should be permitted. One might even persuade the other to change his mind about this normative question (making reference to the trauma suffered by the women in question, or to the rights of the unborn child, or some such), while their meta-ethical differences remain unresolved.

Evolutionary normative ethics says that by studying evolution we will come to an understanding of which ethical principles are the right ones, and which are the wrong ones. As a result of this, the study of evolution might cause us to revise our views about what is right and what is wrong. Evolutionary normative ethics faces an uphill struggle, because it needs to find some way of linking claims about how traits have promoted reproductive success to claims about what is good and bad. This will be a tricky job, for we would certainly be foolish to link the two domains by claiming that whatever tendencies and convictions natural selection promotes are *ipso facto* good. Suppose, for example, that Thornhill and Palmer (2000) turn out to be right about the tendency to rape in human males being an adaptation. This discovery would merely underline the fallacy of inferring that what selection has favoured must be good.

Evolutionary meta-ethics says that by studying evolution we will come to an understanding of the general nature of ethical statements and their subject matter. Proponents of evolutionary meta-ethics might argue, for example, that the study of evolution shows that natural selection has made us believe that there are ethical facts when in reality there are none, or perhaps they might argue that there are ethical facts, and what these facts are is dependent in some way on the evolved nature of our species.

This chapter begins, then, with an overview of Darwin's explanation of the origins of the moral sense. I move on to consider first, evolutionary normative ethics, and second, evolutionary meta-ethics. Darwin invokes the mechanism of group selection in his evolutionary account, and towards the end of the chapter I try to allay the suspicions some might have about the propriety of this mechanism. Finally, I ask whether evolution has made us selfish.

2. THE ORIGINS OF THE MORAL SENSE

In Descent, Darwin steers clear, for the most part, of both evolutionary normative ethics and evolutionary meta-ethics, sticking to providing an explanation of the origins of the moral sense. His account is clearly influenced by the philosophical writings of David Hume and Adam Smith. In broad terms, the story begins with an explanation of how rudimentary moral behaviour, motivated by 'social instincts', is established in humans and animals. These instincts, Darwin says, rely on the action of sympathy, and are laid down by natural selection operating at the level of the community. He then moves on to consider how such things as memory, intelligent reflection and language interact with these instincts to produce a more complex moral sense.

Overall, the account has four main stages (R. Richards 1987). First, Darwin makes a case for the existence of social instincts in animals. These are instincts that lead animals to 'perform various services' for fellow creatures (Descent: 121). Sympathy plays a central role in Darwin's understanding of these social instincts, and of morality in general. Darwin thinks that animals of various kinds feel sympathetically when other creatures are in pain or otherwise distressed; that is to say, one animal feels pain prompted by the pain of another. Sympathy is thus the motivational spur that causes one animal to remove the source of another's pain, thereby extinguishing its own sympathetic pain. Darwin thinks that sympathy between animals is usually restricted to members 'of the same association' – that is, of the same community or family group (ibid.). He explains the prevalence of these sympathetic feelings by a form of natural selection operating between communities: ' . . . for those communities, which included the greatest number of the most sympathetic members, would flourish best, and rear the greatest number of offspring' (ibid.: 130).

Darwin supplements this account of moral behaviours with an account of the origin of conscience. In addition to social instincts, he says, animals have instincts that serve their individual survival. Sometimes the two types of instinct come into conflict. He gives the example of a bird which has an instinctive urge to migrate, thereby securing her own survival against coming cold weather, and a socially instinctive urge to remain where she is and look after her newborn chicks, thereby securing the survival of other creatures. Darwin believes that natural selection gives the social instincts greater 'permanence' than survival instincts. He reminds us that humans, for example, will often reflect ceaselessly and painfully on what others think of them, while the pain of hunger, for example, is quickly gone once satisfied, and is hard to recollect vividly once gone. This means that social instincts will nag at a creature over a long period of time. Even so, the short-term force of survival instincts may sometimes overwhelm social instincts. The bird may indeed migrate, and leave her young to perish. In intelligent organisms endowed with memory and imagination, these cases where social instincts are trumped by survival instincts will lead to very strong negative feelings, for such an organism imagines or perceives the negative social outcome of its actions, and the permanent social instincts are aroused by the sympathetic feelings triggered:

> When arrived at the end of her long journey, and the migratory instinct has ceased to act, what an agony of remorse the bird would feel, if, from being endowed with great mental activity, she

> could not prevent the image constantly passing through her mind,
> of her young ones perishing in the bleak north from cold and
> hunger.
>
> (Ibid.: 137)

This, says Darwin, is conscience: an animal with conscience is one
in which sympathetic pain is aroused by reflection on the consequences of allowing social instincts to go unsatisfied.

The case of the mentally active bird shows that when the social
instincts and more forceful temporary urges come into conflict,
long-lasting misery will often be the result. Darwin believes a kind
of self-command can eventually be acquired that obviates this
misery. Here he draws on the same Lamarckian inheritance mechanism that informs so much of Expression. Imaginative individuals
are able to anticipate the misery that the pangs of conscience will
bring, and they can learn to suppress forceful instincts that might
otherwise undermine social beneficence:

> Man, prompted by his conscience, will through long habit acquire
> such perfect self-command, that his desires and passions will at
> last yield instantly and without a struggle to his social sympathies
> and instincts, including his feeling for the judgment of his fellows.
> The still hungry, or the still revengeful man will not think of
> stealing food, or of wreaking his vengeance.
>
> (Ibid.: 139)

This habit of self-command can be passed to offspring with the
result that, 'at last man comes to feel, through acquired and
perhaps inherited habit, that it is best for him to obey his more
persistent impulses' (ibid.: 140).

We have moved from the first rudimentary stage in which
socially beneficent behaviour is brought about by social instinct,
to the second stage in which conscience and habit interact to
ensure that these social instincts are rarely overwhelmed by self-
serving instincts. The third stage in the origin of the moral
sense marks a partial departure from the social instincts initially
laid down by natural selection. Language now enters the scene,

developing in tandem with the intelligence of the species. Language, Darwin says, coupled with intelligence, makes it possible to formulate and disseminate an explicit common opinion regarding how individuals should act 'for the public good' (ibid.: 122). The role of 'special instincts' – instincts that direct social animals to specific kinds of beneficent behaviour – is now diminished in favour of learned rules, which prompt us to conform with commonly held opinion regarding good conduct.

Instinct is not effaced completely by learning at this stage. Darwin believes that humans are strongly motivated by 'our regard for the approbation and disapprobation of our fellows' (ibid.). Others approve or disapprove of us according to how well our conduct accords with rules that have been communally formulated and endorsed. Hence we act to promote the public good, because we hold the opinion of others in high esteem. Sympathy retains an important role here: Darwin's view is that someone with little sympathetic feeling would not be much concerned by the knowledge that what he did met with the disapproval of others; one must share the onlooker's disapproval, if the onlooker's disapproval is to be effective in modulating one's own behaviour. Thus 'sympathy and instinctive love for his fellows' retain a central role in man's moral motivation even with the development of publicly formulated rules of conduct (ibid.: 132–33).

Darwin does not claim that all publicly formulated rules succeed in promoting the public good:

> The judgment of the community will generally be guided by some rude experience of what is best in the long run for all the members; but this judgment will not rarely err from ignorance and weak powers of reasoning. Hence the strangest customs and superstitions, in complete opposition to the true welfare and happiness of mankind, have become all-powerful throughout the world.
>
> (Ibid.: 146)

He cites a number of publicly endorsed rules of conduct that he regards as irrational: rules of etiquette, Hindu rules of caste, and others (ibid.).

This stage of moral evolution is also marked by an extension of sympathy beyond the family or community. It is basic reasoning, rather than natural selection, which provokes this transition:

> As man advances in civilisation, the simplest reason would tell each individual that he ought to extend his social instincts and sympathies to all the members of the same nation, though personally unknown to him. The point being once reached, there is only an artificial barrier to prevent his sympathies extending to the men of all nations and races.
>
> (Ibid.: 147)

Darwin does not spell out how he thinks reason leads to this conclusion. He might think that reason dictates we extend our sympathies because we realise that the survival of our species depends on the welfare of all peoples worldwide. This seems an unlikely interpretation, because it is hard to square with his claim that the most advanced civilisations extend sympathy to the lower animals. It is more likely that Darwin has in mind either a realisation that it is arbitrary to extend sympathy to those of our community, without extending it to those of other communities who are in no salient ways different from us, or a recognition that the greatest good of the greatest number necessitates the extension of our sympathy.

The final stages in the development of the moral sense occur as increased intellectual faculties, together with a richer basis of experience, allow us to determine the consequences of our actions in greater detail, and eventually to develop the most virtuous habits as a result of that knowledge acting in concert with our extended sympathies. Darwin accepts that such virtuous habits, once acquired through reason and experience, and practiced regularly, may also be inherited in offspring without the need for new learning. This is, in effect, an explanation in terms of use-inheritance for the existence of complex, unlearned moral intuitions. Darwin here quotes at length from the work of Herbert Spencer:

> I believe that the experiences of utility organised and consolidated through all past generations of the human race, have been producing

corresponding modifications, which, by continual transmission and accumulation, have become in us certain faculties of moral intuition— certain emotions responding to right and wrong conduct, which have no apparent basis in the individual experiences of utility.

(From a letter from Spencer to Mill, quoted in *Descent*: 148)

Darwin's story, like Spencer's, is a story of moral improvement in which natural selection plays a diminishing role over time. The natural selection of communities produces creatures with social instincts, motivated by sympathy for members of those communities. Reason and experience lead us, over time, to extend the domain of sympathy 'to men of all races, to the imbecile, maimed, and other useless members of society, and finally to the lower animals' (ibid.: 149). Reason and experience also lead us to a better understanding of which actions foster welfare and happiness in humanity at large. We learn to perform these actions habitually, and eventually they arise automatically in our instinctively virtuous offspring.

3. DARWIN'S NORMATIVE ETHICS

Darwin's story of moral progress is intended to show how selection, and later reason and experience, lead to an improvement in moral conduct over time. This presupposes some view about what makes moral conduct good or bad. Darwin says comparatively little about this. He describes himself as a *utilitarian*, albeit a nonstandard one. Utilitarianism is a particular form of *consequentialism*. Consequentialists believe that the rightness or wrongness of an act depends entirely on its consequences. Utilitarianism, in its classic form, says that the consequences that matter are the net effects of an action on happiness. Classic utilitarianism says that the right action is that which produces the greatest happiness of the greatest number.

It is unclear in what ways Darwin thinks the study of evolution sheds light on the plausibility of utilitarianism. This, in turn, makes it hard to assess whether Darwin backs a form of evolutionary normative ethics. We can appreciate these difficulties by looking at the two modifications Darwin briefly proposes to classic utilitarianism. First, he rejects utilitarianism as a view of 'the motive of conduct', while endorsing it as the 'standard' of conduct (Descent: 144). In other words, utilitarianism correctly diagnoses what makes an action right or wrong, but we should not think that people are generally motivated to maximise happiness. Second, Darwin advocates a revision of the greatest happiness principle as the standard of conduct, and its replacement with a modified principle of the 'general good' (ibid.: 145). He begins with a plausible suggestion, which gives an evolutionary rationale for his scepticism of the greatest happiness principle as the motive of conduct. If social instincts in animals have been formed by selection acting on communities, then these instincts are likely to promote not the overall happiness of the community's members, but their health and strength. For when communities struggle against each other, the health and strength of their members are likely to be more valuable than a cheery disposition. More precisely, Darwin argues that social instincts aim at maximising the 'general good'. This he defines as 'the rearing of the greatest number of individuals in full vigour and health, with all their faculties perfect, under the conditions to which they are subjected' (ibid.: 145). Now Darwin makes the step to a normative conclusion:

> As the social instincts both of man and the lower animals have no doubt been developed by nearly the same steps, it would be advisable, if found practicable, to use the same definition in both cases, and to take as the standard of morality, the general good or welfare of the community, rather than the general happiness.
>
> (Ibid.: 145)

Darwin seems to make an unwarranted leap here from the descriptive claim that natural selection has produced social instincts tending to promote the health and vigour of the community, to the normative claim that the health and vigour of the community ought to be the standard by which we reckon actions good or bad. This is odd, because Darwin is not in the general habit of asserting that whatever selection equips us with is right. He believes that savages who restrict the domain of their sympathy to others in their tribe, and who think nothing of killing their enemies, act wrongly, even though the tendency of selection acting on communities may be to promote such acts.

It is not clear how seriously Darwin wishes to defend his definition of the standard of morality as that which maximises the 'general good', for he immediately adds the cryptic caveat that it 'would perhaps require some limitation on account of political ethics' (ibid.: 145). We can only speculate about what Darwin is alluding to here. Perhaps he is gesturing to worries about how far we should sacrifice the interests of the individual when they conflict with the community. A strict consequentialism that insisted on the 'rearing of the greatest number of individuals in full vigour and health, with all their faculties perfect' might advocate the sterilisation of imperfect specimens, or forced breeding from the strongest in society, or the execution of people with contagious diseases. Perhaps Darwin perceives that his definition will need re-tuning to ensure it does not have these consequences.

The confusion increases when we note that Darwin sometimes appears to distance himself from consequentialism altogether. At the end of chapter four, he summarises what he takes himself to have established:

> I have so lately endeavoured to shew that the social instincts—the prime principle of man's moral constitution—with the aid of active intellectual powers and the effects of habit, naturally lead to the golden rule, 'As ye would that men should do to you, do ye to them likewise', and this lies at the foundation of morality.
>
> (Ibid.: 151)

Darwin seems to say not only that evolution has resulted in humans following the golden rule, but that this rule, which 'lies at the foundation of morality', is the right one for us to follow. Is this compatible with his consequentialist claim that the standard for morality is the 'rearing of the greatest number of individuals in full vigour and health, with all their faculties perfect'? The two principles appear to make quite different demands. Adhering to the consequentialist standard of morality seems compatible with doing to others all kinds of things that ye would rather not that they do unto you, like preventing them from breeding if they are weak. Perhaps Darwin thinks this appearance is illusory, and that the best way to meet the standard of morality is for everyone to follow the golden rule. Perhaps, that is, he thinks that the greatest number will be reared in full vigour and health if everyone acts unto others as they would that others do unto them. Darwin gives no argument for this in *Descent*. This is, however, the kind of proposition that researchers in *evolutionary game theory* have come to investigate in recent years. Roughly speaking, this branch of evolutionary and economic science asks what mixture of behavioural dispositions towards others is likely to emerge in an evolving population of interacting organisms. Research of this sort is relevant to ethics, not merely because it enables us to ask whether a society in which all obey the golden rule is likely, or stable, from an evolutionary perspective, but also because it promises to show whether the empirical assumptions of various ethical and political assertions – 'Society cannot function without strong leaders', 'Welfare will be maximised if each person is left free to pursue his self-interest' – are well-founded.

Let me now summarise what I take to be Darwin's general view of morality. He believes that right actions are those which promote the 'general good'. This is defended as the standard of conduct, not the motive. Yet Darwin believes that, by and large, the actions of individuals do promote the general good. One of the most constructive features of Darwin's evolutionary account is that it helps to square this circle: it allows us to explain why a group of individuals reliably act in ways that promote the general good, in spite of the nature of their conscious motivation. In the first instance this happens because natural selection has equipped us with sympathy and social instincts that promote behaviours beneficial to the community. Later in the evolution of our species, reason and experience cause the domain of sympathy to broaden, and they make us better able to appreciate which types of action tend to the greater good. Darwin portrays man as an animal whose intellectual abilities allow him to recognise the moral inadequacy of the social instincts with which nature initially equips him. Surprisingly for one with such Humean instincts, Darwin includes in this picture the following nod to Kant:

> as love, sympathy and self-command become strengthened by habit, and as the power of reasoning becomes clearer, so that man can value justly the judgments of his fellows, he will feel himself impelled, apart from any transitory pleasure or pain, to certain lines of conduct. He might then declare—not that any barbarian or uncultivated man could thus think—I am the supreme judge of my own conduct, and in the words of Kant, I will not in my own person violate the dignity of humanity.
>
> (*Descent*: 133)

4. EVOLUTIONARY NORMATIVE ETHICS

The primary objection one sees to evolutionary normative ethics is that it makes the mistake of thinking that ethical conclusions can be derived from propositions about evolutionary history. This objection is often re-phrased in a number of ways; sometimes that evolutionary normative ethics ignores the distinction between fact and value, sometimes that it is guilty of deriving an 'ought' from an 'is', sometimes that it commits the 'naturalistic fallacy'. These ways of expressing the stock objection threaten to suck us into a tedious evaluation of details regarding 'ought', 'is', the fact/value distinction and the naturalistic fallacy that is surplus to the requirements of assessing evolutionary normative ethics. Two quick observations regarding these alleged fallacies and distinctions will suffice. First, it is commonly assumed that the error of deriving 'ought' from 'is', and the 'naturalistic fallacy', are the same thing. The latter, many think, is simply a fancy technical name for the former. But the mistake of deriving an 'ought' from an 'is' is one whose fame we owe primarily to David Hume. The naturalistic fallacy was named and diagnosed by G. E. Moore in his much later work *Principia Ethica* (Moore 1903). The concerns of the two philosophers are quite different. Moore's argument is far more recherché than Hume's, and appears to show that no definition of 'the good' can be appropriate, whether in terms of natural or supernatural properties.

Second, many of the technical issues over the inferential gap from 'is' to 'ought' can be ignored for the purposes of our present discussion. There has been a lively debate on this question among philosophers, with some arguing that this gap can sometimes be bridged. Some have suggested, for example, that one can infer merely from the fact that I say, 'I promise to give you £50 on your birthday', that I ought to give you £50 on your birthday. But even if we came to accept that 'ought' could sometimes be inferred from 'is', this would not commit us to endorsing any inferences from evolutionary 'ises' to moral 'oughts'. It is one thing to say that facts about what promises are uttered can yield obligations to the people to whom they are directed. It is another thing entirely to claim that facts about what, in the past, has promoted the ability to reproduce can yield conclusions about what we ought to do. Rather than getting bogged down in technical discussion of 'ought' and 'is', I propose in this section that we look directly at some of the arguments which one might think can link evolutionary facts to normative conclusions.

Let me begin by outlining one way in which evolutionary research might lead us to revise some of our normative claims. We often praise virtues on the grounds that they have some kind of effect or another. Perhaps we think that tolerance is important because it permits the free exchange of ideas, and thereby encourages innovation and progress in society. This reason for valuing tolerance commits us to the view that tolerance really does have these effects. Evolutionary study, when it seeks to shed light on the causes of the proliferation of various virtues, will put these claims to the test. Suppose, for example, evolutionary studies conclude that tolerance spread not because it encouraged innovation, but because groups with tolerators were more easily led and mobilised by charismatic leaders, and as a result defeated other groups in battle. A result like this might lead us to re-appraise tolerance. Note that this is a comparatively weak linkage between evolutionary history and normative evaluation. It reminds us that

some combinations of evolutionary histories and normative evaluations are mutually reinforcing, while others are in tension with
each other. Similar considerations need to be borne in mind in the
context of claims, based on evolutionary game theory, that some
forms of social organisation ought to be rejected because they are
unstable over time. Even if the theorist proves that the social system
in question is evolutionarily unstable, this is only a reason to reject
that system if we are already committed to the ethical importance
of social stability. Since most of us are committed to this – for
without the prospect of future stability there is little point in
making plans for tomorrow, and a life in which no plans are made
is not much of a life at all – work in evolutionary game theory is
indeed relevant to ethical conclusions. But this is not to say that
normative evaluation is entirely dictated by evolutionary facts.

The historian and philosopher Robert Richards has defended a
different form of evolutionary normative ethics (R. Richards 1987:
Appendix II). He begins by observing that when we propose basic
ethical principles, we usually expect to give justifications for them.
Darwin, remember, says that our basic goal should be to maximise
the 'general good', which for him means rearing 'the greatest
number of individuals in full vigour and health, with all their faculties perfect, under the conditions to which they are subjected'
(*Descent*: 145). This claim is not self-evidently true. But how can
we go about justifying it, or any other putative set of basic ethical
principles? Many philosophers think we can do this by showing
how the consequences of basic principles mesh with our intuitions about the rights and wrongs of specific cases. According to
this picture of ethical justification, one would justify Darwin's
view about the general good by showing that cases we would
intuitively regard as terrible – a famine, or war, say – are also
counted as terrible on his view; and by showing that cases we
would intuitively regard as wonderful – a harmonious society of
individuals, all of whom are treated with respect – are also counted
as wonderful on his view. If we wanted to undermine Darwin's
view, we would instead show that it yields results that clash with
our intuitions about right and wrong. We might try to show, for
example, that it tells us we ought to prevent the sick from breeding.

Showing how well an ethical system of goals and principles matches strongly held intuitions is, Richards says, the best way we know to justify an ethical system. But, he adds, our commonly held ethical intuitions are the products of evolution. The evolution of the moral sense endows humans with shared moral intuitions, and those shared moral intuitions are the testing ground for a system of basic moral principles. So this means that an ethical system can, indeed, be justified in virtue of a set of facts about evolutionary history. Richards' argument does not show that we can appeal directly to evolutionary facts in giving justifications – we do not, for example, defend the decision to criminalise zoophilia by trying to show that natural selection has acted against those who have sex with animals. The idea is, instead, that natural selection has led to the proliferation of a strong intuitive aversion to sex with animals, which in turn enables someone who wishes to argue that zoophilia is wrong to appeal to these intuitions in defending their case.

Let me begin by saying what I think Richards has got right. When two people argue, their disagreements are rarely resolved unless they have at least some commonly held beliefs to which they can appeal. It may well be that evolution, by equipping us with many shared convictions about what is right and what is wrong, gives us the shared points of reference we need to engage in productive ethical argument, and to convince each other regarding general ethical principles and specific matters of ethical conduct.

My first problem for Richards is comparatively picky. In what sense do we owe all of our shared ethical beliefs and intuitions to evolution? Darwin, remember, thinks that reason and experience have enabled human societies to formulate publicly expressed moral rules, which most members of those societies learn. If anything like Darwin's story is right, then not all of our shared ethical beliefs and intuitions are explained by natural selection – some are explained by reason, experience and learning. Of course one might reply that the action of reason, experience and learning are all parts of the evolutionary process. We could make this move, but to do so threatens to trivialise 'evolutionary ethics', by making any historical explanation for our shared intuitions an evolutionary one.

A more substantial worry comes from Richards' view that evolution does not merely allow us to make convincing moral arguments, it enables us to justify our moral principles. But is being convinced of a view the same as being justified in holding that view? The analogy with science suggests not. If a group of scientists – intelligent design theorists, say – share a lunatic set of principles about what constitutes good evidence, then I might be able to convince those scientists of some crackpot belief by trading on their principles. That belief will not be justified, because their principles, although shared, are bad principles. So even if evolution has enabled humans to formulate ethical arguments that are convincing to other members of the species, that does not mean that evolution makes our ethical arguments justified.

Richards hints at a possible response to this when he summarises his stance on ethical justification. He says that 'frameworks, their inference rules, and their principles are usually justified in terms of intuitively clear cases—that is, in terms of matter of fact' (Richards 1987: 617). This equation of 'intuitively clear cases' with 'matter of fact' is ambiguous. Perhaps Richards is saying that anything we find to be intuitively clear is a fact. We have a strong intuition that murder is wrong; hence it is a fact that murder is wrong. Strong ethical intuitions should be treated as individual pieces of ethical knowledge – as data points – which the principles of ethical theory should seek to systematise. The analogies Richards draws between ethical and scientific justification invite this understanding of his claims. But the analogy with science reminds us that we would not normally endorse the identification of what many people intuitively believe with what is fact. Of course Richards might say that we have misread his remark. He means to say only that it is a fact that we have many shared ethical intuitions. But once again this brings us back to the question of how shared intuitions can justify an ethical theory. Could not natural selection have equipped us with mistaken moral intuitions? Darwin seems to accept such a possibility when he claims that good reason tells us to broaden the domain of sympathy beyond that which natural selection builds into us. Of course ethics might be fundamentally different from science; perhaps in the realm of ethics there is nothing more to being a fact than being intuitively believed by many people. But this would need arguing for, and such arguments bring us to the subject of meta-ethics.

5. EVOLUTIONARY META-ETHICS

If Darwin's historical account of the evolution of the moral sense is right, then an alternative historical account – namely that God planted the moral sense in our minds – is wrong. Emma Darwin appeared worried that this might undermine the authority of morality. She once told her son Francis, 'Your father's opinion that all morality has grown up by evolution is painful to me' (quoted by Brooke 2003: 202). Is an evolutionary history for morality also a trivialising one? If the only motivation one might have for acting morally is the thought that God put the moral sense into one's head, then things look bad. But Darwin's story, as we have seen, is framed as one in which evolution brings progress to human virtue. Could one perhaps say that evolution brings us to closer knowledge of the moral law?

The biologist E. O. Wilson and the philosopher Michael Ruse have together argued that Darwin's work should push us towards a form of scepticism about morality (Ruse and Wilson 1993). They argue that natural selection can explain why humans tend to believe in a realm of objective moral fact. Natural selection explains, for example, why humans tend to believe it is a fact that murder is wrong. Once we appreciate natural selection's power to account for our moral convictions, we no longer have any reason to posit the existence of a realm of moral fact to explain these shared moral convictions. Ruse and Wilson say that the study of evolution shows us that 'human beings function better if they are deceived by their genes into thinking that there is a disinterested objective morality binding upon them, which all should obey' (ibid.: 425). They argue that we 'think morally because we are subject to appropriate epigenetic rules', which 'give the illusion of objectivity to morality' (ibid.: 426–27). So Ruse and Wilson hold that evolution has given humans a set of false convictions in the existence of an objective morality. The job for this section is to see whether the study of evolution really has such strong meta-ethical implications. Is it true, for example, that someone who believes that murder is objectively wrong has been deluded by natural selection?

Ruse and Wilson's picture is complicated by their closing remarks, in which they suggest that only by studying evolution 'will we see how our short-term moral insights fail our long-term needs, and how correctives can be applied to formulate more enduring codes' (ibid.: 436). This is consistent with the view that morality is an illusion, but it implies that not all moral illusions are equal. The study of evolution, they say, will tell us which illusions are best for our species in the long-run. Ruse and Wilson add an additional layer of difficulty to their view when they espouse a kind of moral relativism:

> No abstract moral principles exist outside the particular nature of individual species. It is thus entirely correct to say that ethical laws can be changed, at the deepest level, by genetic evolution. This is obviously quite inconsistent with the notion of morality as a set of objective, eternal verities. Morality is rooted in contingent human nature, through and through.
>
> (Ibid.: 431)

Once again, we can make this consistent with their other claims. All of morality is an illusion, but the question of which illusions it is best for a species to adopt depends on the nature of that species.

I want to focus on Ruse and Wilson's basic claim that 'objective morality' is an illusion. Scepticism of objective moral facts has a long pedigree in philosophy. It is not hard to see why. What kinds of things could moral facts be? Where are they? How on earth do we find out about them? More to the point, what difference would they make? Perhaps we could appeal to them in settling arguments about what we ought to do. But a homicidal maniac might still say, 'Even though you have convinced me that it is a fact that murder is wrong, such facts are nothing to me. I still choose to murder'. Facts do not seem to have the motivational force that might help us convince a moral monster to change their ways. The philosopher J. L. Mackie famously said that moral facts, if they existed, would have to be rather 'queer' things; so queer that we probably shouldn't admit their existence at all. Many philosophers have followed David Hume in adopting a picture of moral evaluations as projections of the human mind onto a world that has no moral facts. So we need to be careful about what question we ask. The issue is not whether objective morality is an illusion. The issue is whether the study of evolution gives us any new reasons for believing objective morality is an illusion.

Let us grant Ruse and Wilson the claim that natural selection explains why we have many of the convictions we do, moral or otherwise. By itself, this does not show that any of our convictions are false. Plausibly, natural selection explains why we have the conviction that snakes are dangerous. But we should not think this conviction is false because of this.

Why, then, say that if we can show that ethical convictions have evolutionary histories, we thereby show that they are false? Perhaps Ruse and Wilson think the answer lies in an important difference between convictions about snakes being dangerous and convictions about stealing being wrong. We can tell an evolutionary story in which the fact of snakes being dangerous explains why the conviction that snakes are dangerous evolved. People who believed that snakes were not dangerous got bitten by them and died; people who believed that snakes were dangerous avoided death. It is far from clear how to tell a similar story for ethical convictions. Such a story would need to show how the fact of stealing being wrong explains why people who believed that stealing was wrong enjoyed greater chances of survival and reproduction than people who did not.

So does this mean that evolution shows ethical claims to be false? It does not. We have particular difficulty in telling an evolutionary story which shows how the fact of stealing being wrong has a favourable causal impact on the reproductive success of people who believe it. But the first point to note is that this problem is not new; it merely reflects traditional philosophical worries about what kinds of things moral facts could be, and how they could interact with us (Sober 1994a; Kitcher 1993).

Second, even if we do not interact causally with moral facts, there might still be such facts, and natural selection might still reliably cause us to believe them. Suppose it is a fact that it will rain tomorrow. We can believe this truly now not by interacting causally with tomorrow's rain, but by making inferences from different facts that we have causal access to today, so long as these facts are correlated in the right way with tomorrow's weather (facts about barometers, weather forecasts, and such like). Similarly, if there are non-moral facts which interact with us causally, and which correlate in some way with moral facts, then natural selection might lead us to believe in moral facts.

Third, Ruse and Wilson's argument threatens to show that mathematics, as well as morality, is an illusion. We tend to think it is an objective truth that $2+2 = 4$. What is more, we might be able to give an evolutionary explanation for some of our more basic mathematical beliefs. Perhaps a rudimentary ability to add up aids survival and reproduction. However, it is not clear what kind of fact $2+2 = 4$ is. Where is it? Does it cause people who believe it to have greater reproductive success than those who do not? If not, how might natural selection explain how we come to believe it truly? It would be hasty, to put it mildly, to jump from the difficulty we have in giving an evolutionary explanation of this sort to the conclusion that mathematics is an illusion foisted on us by our genes. If we should not make this jump for mathematics, we should not make it for morality either.

Fourth, and finally, we should not be too quick to dismiss the possibility of constructing a selection explanation that links moral facts to moral convictions. It all depends on what kinds of things moral facts are. Suppose, for example, that moral facts are facts about what is in the community's interests. If group selection is efficacious, then people will tend, over time, to act in ways that promote the cohesion and vigour of their communities. Just as individual selection tunes people's beliefs to the true danger of snakes, so group selection tunes people's beliefs to the true interests of the community. The philosopher Peter Railton uses an argument like this to make a case for seeing natural selection as a feedback mechanism, which explains why individuals believe moral facts (Railton 1986).

Ruse and Wilson say that an evolutionary explanation for our moral beliefs 'makes the objective morality redundant, for even if external ethical premises did not exist, we would go on thinking about right and wrong in the way that we do' (Ruse and Wilson 1993: 431). If, however, objective moral facts are facts about the interests of the community, then these facts are not redundant after all. If the moral facts had been different, then the community's interests would have been different, and group selection would have tuned our moral sense to reflect that. Of course one will respond by saying I have given no reason to think that moral facts are facts about what is in the community's interest. Indeed I have not, but I raise this argument to show not that moral realism – the view that there are objective moral facts – is the correct position, but to suggest a way in which moral realism can make room for evolution.

6. GROUP SELECTION

Darwin's history of the moral sense draws on a form of selection that some of today's biologists view with suspicion. This is selection at the level of the community or, as we would now say, 'group selection'. Darwin invokes community selection to solve a problem in the explanation of social instincts and sympathy. This problem is a special instance of the more general *problem of altruism*. It is easiest to appreciate what this problem is with an example. Animals sometimes patrol their territories, and give warnings when predators come near. This patrolling behaviour benefits others in the group, but it also has a cost, which falls disproportionately on the patroller, who risks being the first to be caught and eaten by the predator. Patrolling is an example of an altruistic behaviour, because patrolling makes selfish non-patrollers, who get the benefit without paying the cost, fitter than patrollers themselves. In general, altruistic behaviours are those which reduce the individual fitness of the actor relative to other organisms. Our definition entails that altruistic individuals are always less fit than selfish individuals. This, in turn, suggests that natural selection will always favour selfishness. Darwin recognises this problem very clearly in the specific context of sympathy:

how, within the limits of the same tribe did a large number of
members first become endowed with these social and moral qualities . . . ? It is extremely doubtful whether the offspring of the
more sympathetic and benevolent parents, or of those who were
most faithful to their comrades, would be reared in greater numbers than the children of selfish and treacherous parents
belonging to the same tribe. He who was ready to sacrifice his
life, as many a savage has been, rather than betray his comrades, would often leave no offspring to inherit his noble nature.
The bravest men, who were always willing to come to the front in
war, and who freely risked their lives for others, would on average
perish in larger numbers than other men.

(*Descent*: 155–56)

If sympathy and benevolence always have a net fitness cost to the
individual, how can they evolve? Darwin's response to this
problem is to argue that selection can occur at levels above the
individual. So although altruistic individuals who are sympathetic,
who help others, and so forth, will do worse than selfish individuals within the tribe, tribes with large numbers of selfish
individuals will perish at the hands of tribes with altruists:

It must not be forgotten that although a high standard of morality gives
but a slight or no advantage to each individual man and his children over
the other men of the same tribe, yet that an increase in the number
of well-endowed men and an advancement in the standard of morality
will certainly give an immense advantage to one tribe over another.

(Ibid.: 157)

Group selection of this kind is not impossible, but some modern
biologists have claimed that the problem of 'subversion from within'
makes it inefficacious in practice (Williams 1966). Consider again
the case of patrolling. Let us suppose that there are two types of creatures: call them 'patrollers' and 'loafers'. Groups with high
proportions of patrollers will do better in terms of offspring
numbers than groups with high proportions of loafers. But when
we look within a group, regardless of what kind of group it is, we

will find that the loafers out-compete the patrollers. It seems likely, then, that even a group with a high proportion of patrollers will, over time, become overrun with loafers – it will be 'subverted from within'.

This is not the place to explore the literature on group selection in any detail, but it is worth saying a little about the extent to which Darwin's views on the origins of the moral sense are compromised by the scepticism that many hold about group selection. The first thing to note is not all of today's biological thinkers are so set against it. Some theorists – most notably Elliott Sober and David Sloan Wilson (Sober and Wilson 1998) – have disputed the claim that individual selection will always swamp group selection. They have not convinced all of their critics, but their arguments are strong, and they deserve the most serious attention.

One often hears the claim these days that the right way to explain the evolution of altruistic behaviour is not to shift up to the level of the group, but down, to the level of the gene. At first sight, this is a curious assertion. If the problem for group selection is disruptive subversion from the level of individual organisms below, then individual selection should be equally troubled by subversion from an even lower level – that of the gene. If selection goes on at the level of the gene, then surely well-adapted individual organisms will be torn apart by competition among the genes that make them up, just as well-adapted groups are torn apart by competition among the individuals that make them up?

It turns out that something rather misleading is going on when one speaks of solving the problem of altruism by 'shifting down a level' to look at genes. The relationship between the level of the gene and the level of the individual is not, in fact, analogous to the relationship between the level of the individual and the level of the group. Groups are liable to subversion because the individual organisms within a given group can have different numbers of offspring. If the individual genes making up a single organism also had different numbers of offspring, then the organism would be liable to subversion, too. When we are advised to explain altruism by looking to the level of the gene, are we being asked to consider competition between genes in this sense?

Let us look at how a gene-selection view might make sense of the evolution of patrolling. Imagine, for the sake of argument, a population of organisms whose members reproduce asexually. Suppose these organisms have one of two genes. Gene P produces patrolling, gene L produces loafing. Group 1 is a family group, most of which have P, but a few have L. Group 2 is a different family, most of which have L, but a few have P. Because Group 1 is blessed with so many patrollers its members will have many more offspring than those of Group 2, who are more vulnerable to predators. What is more, the majority of the offspring produced by Group 1 will be patrollers. Within Group 1, each individual with L has more offspring than each individual with P. The same goes for Group 2. But when we look at both families combined, the frequency of gene P, and consequently the frequency of patrolling, increases. This happens because the loafers in Group 2, who suffer greatly from predation, do not produce enough loafing offspring to compensate for the large numbers of patrollers being spewed out by the well-protected Group 1.

Suppose the moral we draw from this story is that we should explain the evolution of patrolling by reference to the greater fitness of gene P compared with gene L. What are we saying here? We are not asserting the existence of different reproductive rates of individual genes within each single organism. Rather, we are asserting that when we look across both groups, the patrolling effect of P causes genes of type P to be produced in greater proportions than genes of type L. Individual selection threatens to subvert group selection because individuals of different types have different numbers of offspring within a single group. Gene selection does not threaten to subvert both group and individual selection, because when people talk about gene selection as an explanation of altruism, they are not asserting the existence of genes of different types that have different numbers of offspring within a single individual.

This all goes part of the way to substantiating an apparently shocking claim made by Sober and Wilson. They make a lot of the efficacy of group selection in their explanation of altruism. Are they saying that gene-level selection is not important? They reply that gene-level models for the evolution of altruism are, in fact, group selection models in disguise (e. g. Sober and Wilson 1998: 77). We can see why they say this. The model we just looked at explains the evolution of patrolling by reference to the greater fitness of P genes compared with L genes. But it is also a model in which groups with lots of patrollers have many more offspring than groups with few patrollers. Patrolling thereby increases its frequency in the population considered as a whole, even though within groups, individual patrollers are always less fit than individual loafers. This makes the model look like a group selection model after all.

If Sober and Wilson are right, then the current success of gene-level models for the explanation of altruism may, in fact, support group selection of the kind that Darwin advocates. Sober and Wilson have W. D. Hamilton as an ally here. Hamilton invented 'kin selection', which is one of the gene-level models Richard Dawkins took himself to be popularising when he published The Selfish Gene. Dawkins's book is, of course, the Bible of gene-level selection. But Hamilton agrees with Sober and Wilson that kin selection is a form of group selection (Okasha 2001: 25).

7. HAS EVOLUTION MADE US SELFISH?

The topic of selfishness and altruism merits more attention, and more care, than we have given it so far. Michael Ghiselin has written: 'If the hypothesis of natural selection is both sufficient and true, it is impossible for a genuinely disinterested or "altruistic" behavior pattern to evolve' (Ghiselin 1973: 967).

I do not think this was Darwin's view, nor do I think it is true. But to see why natural selection can produce 'genuinely disinterested' behaviours we first need to get clear on what we mean by 'altruism'.

Let us begin by distinguishing 'psychological altruism' and 'biological altruism' (Sober 1994b). When we call a person selfish or altruistic we are often saying something about their motivation. A selfish person is someone who cares only about themselves; an altruist cares about others. Selfish people, for example, might give heavily to charity only because it gets them invited to swanky dos where they can lobby important politicians informally, and make even more money for themselves. Altruists, on the other hand, give heavily to charity because they care about the poor, or the starving. To make a claim about altruism or selfishness is here to make a psychological claim.

Biological altruism is an entirely different matter. Biologically altruistic behaviours are classed as such not in virtue of their motivational causes, but in virtue of their effects on fitness. We already gave a definition of a biologically altruistic behaviour in our discussion of patrolling. A biologically altruistic behaviour is one that reduces the fitness of the actor relative to the recipient. So patrolling is a biologically altruistic act because even if the actor gains a benefit, the recipient gains the benefit without paying the price.

There can be biologically selfish acts that are psychologically altruistic. Donating semen to a sperm bank because one wants to help infertile couples is an example. Conversely, there can be psychologically selfish acts that are biologically altruistic. Giving money to charity in order to increase one's social standing might be an example, if increased social standing does not translate into reproductive success.

If selection acts only at the level of the individual then, as we have seen, it is impossible for biological altruism to evolve. It is part of the definition of biological altruism that altruistic individuals are less fit than selfish ones. I argued in the last section that Darwin's group selection explanation for the emergence of biological altruism deserves to be taken seriously. But even if we were to accept the impossibility of biological altruism, that would not commit us to dismissing psychological altruism. Natural selection might have made use of psychologically altruistic mechanisms to bring about biologically selfish behaviours.

Consider mothers who look after their offspring. Caring for offspring is a biologically selfish behaviour. Perhaps this behaviour is caused by a psychologically selfish mechanism. Perhaps mothers care only about their own happiness, and they know that if their offspring are injured this will make them unhappy. Caring behaviour could also be brought about by a psychologically altruistic mechanism. Perhaps mothers care not for their own happiness and welfare, but for the welfare of their offspring. Perhaps they are motivated to act when their offspring's welfare is put in jeopardy.

Is natural selection more likely to favour psychological selfishness, or altruism? The psychologically altruistic mother may in fact be fitter than the psychologically selfish mother. One might expect strong viable offspring to be reared more effectively by a mother who is directly concerned for the welfare of those offspring, than by a mother who is primarily concerned about her own welfare, and who assists her offspring only in so far as she calculates that this will promote her welfare (Sober 1994b). Of course the matter will not be settled by speculation, but by empirical investigation. Here is another area in which detailed evolutionary studies can have a real impact upon ethics. Darwin's own view, for what it is worth, is that natural selection has equipped us with many social instincts that prompt beneficent action with no prior calculation of the pain or pleasure they bring:

> . . . many a civilised man, or even boy, . . . has disregarded the instinct of self-preservation, and plunged at once into a torrent to save a drowning man, though a stranger. . . . Such actions . . . are performed too instantaneously for reflection, or for pleasure or pain to be felt at the time.
>
> (*Descent*: 134)

One might think that actions like these that involve no reflection whatsoever are mere reflexes, and as such fall outside the scope of moral evaluation. Darwin denies this. But his comments show that he is by no means wedded to the view that we are all fundamentally selfish, either biologically or psychologically speaking.

SUMMARY

Darwin's interests in evolution and ethics focus in the main on an attempt to give an historical explanation for the emergence of our moral sense – the sense of what is right and what is wrong. His story is one of moral progress, driven by group selection. The struggle for survival between communities initially favours those communities whose members feel sympathy for each other, and who are consequently motivated to act in ways that strengthen the community as a whole. As time goes by, experience and reason cause the domain of sympathy to expand, encompassing humans from other communities, and eventually non-human animals. We also gain the ability to formulate, disseminate and enforce rules of good conduct generated by reason and experience. In these respects Darwin's account of the evolution of the moral sense is rather like his account of the evolution of the emotions: natural selection is part of the story, but by no means the only part of that story. By and large, Darwin is careful not to overstate the significance of an evolutionary account of the origins of the moral sense. This is just as well, for while several theorists have made strong claims about how an evolutionary account might change our view of morality, it is very hard to substantiate such claims. For example, we saw little hope for attempts to draw conclusions about what we ought to do solely from premises about evolutionary history. We also saw that an evolutionary account of the origins of the moral sense need not mean that we are all selfish, and it does not show that there are no moral facts.

FURTHER READING

Darwin discusses the evolution of the moral sense in chapters four and five of *Descent*.

Robert Richards discusses Darwin's ethical thought in detail:

Richards, R. (1987) *Darwin and the Emergence of Evolutionary Theories of Mind and Behavior*, Chicago: University of Chicago Press.

Two very useful papers clarifying a number of issues regarding the links between evolution and ethics, and to which I owe many of the arguments of this chapter, are:

Kitcher, P, (1994) 'Four Ways of "Biologicizing" Ethics', in E. Sober (ed.) *Conceptual Issues in Evolutionary Biology*, second edition. Cambridge, MA: MIT Press.

Sober, E. (1994a) 'Prospects for an Evolutionary Ethics', in E. Sober, From a Biological Point of View, Cambridge: Cambridge University Press.

Kitcher and Sober's articles are responses to the evolutionary ethics defended by Ruse and E. O. Wilson in:

Ruse, M. and Wilson, E. O. (1993) 'Moral Philosophy as Applied Science', in E. Sober (ed.) Conceptual Issues in Evolutionary Biology, second edition, Cambridge, MA: MIT Press.

Elliott Sober picks over the relationship between evolution and selfishness here:

Sober, E. (1994b) 'Did Evolution make us Psychological Egoists?', in E. Sober, From a Biological Point of View, Cambridge: Cambridge University Press.

Sober and D. S. Wilson defend their views on group selection in an important work:

Sober, E. and Wilson, D. S. (1998) Unto Others: The Evolution and Psychology of Unselfish Behavior, Cambridge MA: Harvard University Press.

Evolutionary game theory is a fecund area for work on ethics, and one which is only mentioned briefly in this chapter. Two important and accessible works in this domain, which are also philosophically engaged, are:

Skyrms, B. (1996) Evolution of the Social Contract, Cambridge: Cambridge University Press.
Binmore, K. (2005) Natural Justice, Oxford: Oxford University Press.

Seven

Knowledge

I. WHAT IS KNOWLEDGE?

What is it for a person to know something? Questions like this one make up the subject matter of *epistemology*, the philosophical study of knowledge. Giving a detailed answer to our question is tricky, but there are some general points we can be fairly sure of. Consider Tony. He might *say* he knows there are weapons of mass destruction in Iraq, but he does not know this unless there *are* weapons of mass destruction in Iraq. Tony's case suggests that knowledge requires true belief, but true belief is not enough for knowledge. Suppose that Nick is convinced that Sunderland won yesterday's football match. And they did win. But Nick believes this because he glanced at the 'Mighty Sunderland Triumphant!' headline from last week's paper, thinking it was today's. Nick's belief is true, but its truth is, in some sense, an accident; he has been lucky. So knowledge is true belief, whose truth is non-accidental.

There are two ways in which Darwin enters into epistemology. In the second half of this chapter we will look at the increasingly popular conception of the growth of scientific knowledge as an evolutionary process, in which different theories compete with each other, the better adapted theory winning out. This 'evolutionary epistemology' takes Darwin's views regarding competition between organisms and applies them to wholly different kinds of entities – scientific theories. Although Darwin himself does not try to show that scientific change is a selection process, he does argue that languages evolve as a result of non-biological selection. He therefore gives an early defence of the claim – endorsed by evolutionary epistemologists – that evolutionary processes are not restricted to the organic realm. In *Descent* he develops the view (mentioned earlier in *Origin*) that the basic mechanisms of change are the same, whether one is talking about linguistic change or organic change. He quotes the linguist Max Müller with approval:

> A struggle for life is constantly going on amongst the words and grammatical forms in each language. The better, the shorter, the easier forms are constantly gaining the upper hand, and they owe their success to their own inherent value.
>
> (*Descent*: 113)

Darwin adds that 'mere novelty and fashion', rather than inherent value, may sometimes be responsible for the preservation of some words, and he states his view explicitly that linguistic change does not merely show loose analogies with organic change, but rather: 'The survival or preservation of certain favoured words in the struggle for existence is natural selection' (ibid.).

There is a more direct application of Darwin's ideas to epistemology, which will be our concern for the first half of the chapter. Many scientists are convinced that some proportion of human knowledge – whether that is knowledge-how or knowledge-that – is innate. To say that a person knows something innately is to say, at a minimum, that learning does not explain their knowledge. But if learning is not the explanation for the possession of a true belief (or the possession of some valuable ability), what is? Plato explained innate knowledge by claiming that we remember what our souls knew before we were born. Answers like this were, unsurprisingly, unpalatable to many pre-Darwinian philosophers, and they were consequently sceptical of the existence of innate knowledge. Darwin suggests a more plausible mechanism for the generation of innate knowledge: a person can know something that his own experience cannot explain, but that inheritance from his ancestors can. As he remarks in his M notebook: 'Plato . . . says in *Phaedo* that our "*necessary ideas*" arise from the preexistence of the soul, are not derivable from experience.— read monkeys for preexistence' (quoted in Barrett et al. 1987: 551).

2. EMPIRICISM

Empiricism – the school of philosophical thought in the tradition of Locke, Berkeley, Hume and others – gives a leading role to experience in the acquisition of knowledge. As a consequence, empiricists are traditionally sceptical of appeals to innate knowledge. Philosophically, Darwin had strong empiricist inclinations: he admired the work of David Hume, and he regularly expressed frustration with those who speculated on metaphysical questions without attempting to relate these questions to observational evidence. On the other hand, Darwin was an enthusiastic advocate of innateness, especially in his later years. In his autobiographical reminiscences he writes: 'I am inclined to agree with Francis Galton in believing that education and environment produce only a small effect on the mind of any one, and that most of our qualities are innate' (Autobiography: 20).

It appears that this was not always his view, and that Darwin changed his mind after reading Galton's book on the inherited nature of intelligence, Hereditary Genius. He wrote to congratulate his cousin on the work:

> I do not think I ever in all my life read anything more interesting and original. And how well and clearly you put every point! . . . You have made a convert of an opponent in one sense, for I have always maintained that, excepting fools, men did not differ much in intellect, only in zeal and hard work . . .
>
> (Darwin 1903: 41)

Putting things simply to begin with, we can contrast the view held by the older Darwin, which claims that some significant proportion of the facts we know and the skills we have (our knowledge-that and our knowledge-how) is innate, with that of the radical empiricist camp, which says that all beliefs and skills are learned. Darwin challenges the radical empiricist picture in two ways. We have seen countless times that Darwin believed that a habit, learned and practised during the life of an individual, could be passed on to the individual's offspring in such a way that no re-learning of the habit would be required. This mechanism of 'use-inheritance' challenges the claim that all knowledge an individual holds must be acquired by learning during the life of that same individual. But it does not challenge the claim that all knowledge must be acquired by learning during the life of someone or another: use-inheritance respects the empiricist principle that all knowledge ultimately has its source in individual learning.

Natural selection challenges radical empiricism more fundamentally than does use-inheritance, for it suggests a plausible mechanism whereby true beliefs can be reliably acquired without any individual learning. If beliefs are inherited, if there is variation with respect to which beliefs individuals have, and if true beliefs are beneficial in the struggle for life, then over time the members of a population could come to have a number of beliefs that are not true by accident, even though learning is not the process that explains their truth. At a stretch one could say that the population as a whole has 'learned' by a kind of trial-and-error process, and, of course, the knowledge the population acquires relies on individuals interacting with their environment in such a way that those with beliefs closer to the truth are more successful than those whose beliefs are more erroneous. So although natural selection challenges the empiricist regarding the importance of individual learning, the more general empiricist principle that knowledge must be acquired by interaction with the world remains firm, as does the principle that knowledge requires learning in some extended sense of that term. In this way, Darwin appears to make innateness a respectable concept for the more moderate empiricist.

3. INNATE KNOWLEDGE

Let us investigate the assertion that natural selection renders innate knowledge respectable in a little more detail. When Darwin gives evidence for his view that many of his own mental traits are innate, he does not cite the similarities between himself and his siblings as one might expect, but the differences: 'The passion for collecting, which leads a man to be a systematic naturalist, a virtuoso or a miser, was very strong in me, and was clearly innate, as none of my sisters or brothers ever had this taste' (*Autobiography*: 7).

Darwin supposes that he and his siblings, because they have been brought up in the same household, have had roughly similar upbringings. Any differences between them are therefore most likely due to different innate endowments. Darwin's argument expresses the thought that innate traits are skills, dispositions or beliefs that are not learned. I will express some scepticism about the concept of innateness at the end of this section. For the moment, let us follow Darwin in assuming that the equation of 'innate' with 'unlearned' is not overly problematic.

Two questions immediately present themselves. First, what is the relationship between truth and fitness? Might innate errors, rather than innate knowledge, sometimes be more useful in the struggle for existence? Second, what is the relationship between innateness and fitness? Should we expect natural selection to equip organisms with innate beliefs, or with a capacity to learn?

Let us tackle the first question by looking to Friedrich Nietzsche, a philosopher whose explicit comments about Darwin are all negative. In spite of his avowed hostility, Nietzsche's remarks on the 'Origin of Knowledge' in *The Gay Science* begin with a crisp statement of a Darwinian epistemology:

> Over immense periods of time the intellect produced nothing but errors. A few of these proved to be useful and helped to preserve the species: those who hit upon or inherited these had better luck in their struggle for themselves and their progeny.
>
> (Nietzsche 1974: section 110)

However, for Nietzsche, evolutionary epistemology argues in favour of the falsehood of many of our beliefs:

> Such erroneous articles of faith, which were continually inherited, until they became almost part of the basic endowment of the

species, include the following: that there are equal things, sub-
stances, bodies; that a thing is what it appears to be; that our will
is free; that what is good for me is also good in itself.

(Ibid.)

Here Nietzsche is laying down a challenge: why suppose truth
more adaptive than falsehood? As he later puts it, 'The conditions
of life might include error' (ibid.: section 121).

Nietzsche is interested in grand metaphysical concepts; free
will, causation, identity. A more down-to-earth example shows
how under some circumstances a creature may be better off
innately believing what is false, than learning to believe what is
true (Godfrey-Smith 1996; Sober 1994c). Consider two strategies
for determining an animal's behaviour towards snakes. It might be
'hard-wired' to believe all snakes are dangerous. On the other
hand, it might inspect each snake in turn in order to determine
how dangerous it is. The first strategy is highly error-prone.
Specifically, it is prone to so-called 'false positives': all dangerous
snakes will be correctly recognised, but it classifies as dangerous
many snakes which are harmless. The second strategy may be
more accurate, all things considered. It might result in far fewer
false positives. But this second strategy is more prone than the first
to 'false negatives' – on occasions it will yield the result that a
particular snake is not dangerous when in fact it is. This fact can
make the first strategy fitter than the second, because the costs of
false positives and false negatives are not identical. It is much
worse to approach a snake in the false belief that it is safe, than it
is to run away from a snake in the false belief that it is dangerous.
Of course if there are thousands of snakes around, of which
almost none are poisonous, the innate falsehood may in fact be
more costly than learning, simply because it so frequently causes
unnecessary panic. The details of these models are interesting, and
depend in quite subtle ways on the specifics of local circumstances. But in general, if the costs of false negatives are high, then
a false innate belief can be fitter than a learning mechanism whose
overall accuracy is far greater, but which is slightly more prone to
false negatives. So although it is possible that natural selection

gives us innate beliefs that are also knowledge, this is by no means assured in all cases.

There are additional considerations to take into account, which suggest that a mixed bag of innate beliefs and learning strategies is likely to evolve. Learning is sometimes time-consuming, it is not perfectly reliable, and it also demands considerable resources of energy. Of course it may be too hasty to assume that 'hard-wired' beliefs, which must also develop in the maturing organism, are less energetically demanding, and more developmentally dependable, than learned beliefs. Assuming – and it is a big assumption – that 'hard-wiring' is cheaper than learning, it seems that in an environment that changes slowly or not at all, creatures that acquire skills or beliefs by learning may be at a disadvantage compared with creatures in which those same skills or beliefs are 'hard-wired'. But this does not mean that learning is never favoured. In changing environments, creatures with 'hard-wired' skills or beliefs may find themselves developing traits that would have worked in an old environment, but are now a liability.

How does our own species fit in to this? Our physical environment probably changed quickly, even during the extended hunter-gatherer period of the Pleistocene. Climatic conditions, for example, were not constant during the epoch that stretched from 1.8 million to 10,000 years ago. One might think variability of this kind is largely irrelevant, for many of the most pressing problems faced by our species concerned negotiating the social environment, not the physical environment. At a broad level of analysis many social problems may have been quite stable for our ancestors. Perhaps it is always good to repel attacks from enemy bands. Perhaps it is always good for females to choose the fittest males as mates. But these problems lose their stability when we look at them in more detail (Buller 2005: 99). If enemy bands are able to invent new ways of killing and injuring, then it is no use having innate skills for defending against the last generation's karate chop. If males are good at faking their fitness, then a girl sets herself up for disappointment if she uses innately specified criteria that no longer work to sort the genetic wheat from the chaff. At this finer level of analysis, our social environment, too, was probably subject to rapid change. So once again, although it is possible for natural selection to build innate beliefs into a species, there are reasons to doubt that this is always the best adaptive strategy, and there are reasons to think it may not have been the best response to many of the problems faced by our ancestors (Sterelny 2003a).

Let me briefly relate these considerations to one of the Santa Barbara School's tenets, which I left dangling in chapter five. This group of evolutionary psychologists holds that the mind is 'massively modular'. The mind, they say, is composed of many specialised cognitive mechanisms, each of which evolved to solve some specific evolutionary problem, such as the evaluation of potential mates, or the detection of social 'free-riders'. In general, specialised mechanisms are usually better at solving problems than general ones. That is why you have so many different gadgets in your kitchen, instead of using one spoon for reaming lemons, balling melons, beating eggs and crushing garlic. The Santa Barbara School argues that a mind composed of many special-purpose modules will therefore be fitter than a mind composed of a few general-purpose ones. But modules, they say, are also innate.

The arguments we have looked at so far suggest it is a mistake to pit the Santa Barbara School's enthusiasm for innateness against the traditional empiricist's scepticism of innate knowledge. The Santa Barbara School argues primarily that the cognitive capacities that enable us to acquire and process information are innate. They rarely argue that beliefs – pieces of information themselves – are innate. In asserting the existence of innate modules the Santa Barbara School does not deny that many of our beliefs are learned, but it holds that we have various innately-specified cognitive structures, which affect how we learn and process information. Conversely, those who argue that learning is the source of all knowledge – 'blank-slaters' in the empiricist tradition – need not claim that the mind has no innate structure, only that there is no innate knowledge.

In fact, one might think that the mind must have some innate structure. For everyone will agree that the capacity to learn must develop somehow in the embryo. Moreover, everyone will agree that learning cannot be learned, for how could the developing human come to learn by using a capacity that by hypothesis it does not yet have? So if we understand 'innate' as 'non-learned', everyone must agree that at least one of our cognitive capacities – namely the ability to learn itself – is innate.

This argument does not really establish the necessity of innate cognitive structure; rather, it highlights the inadequacies of our definition of 'innate' as 'non-learned'. If we are debating how beliefs are acquired, it may be satisfactory to use 'innate' to mean 'non-learned'. Even in this context the definition has its problems, for learning itself can be defined in different ways, and depending on how demanding our definition is we are likely to count very many, or very few, beliefs as innate. But once we shift to enquiring about the innateness of traits for which learning is not even a candidate mode of acquisition, a new definition of 'innate' is required for our questions to make sense. My scars did not arise through a learning process, but that does not make them innate. The fact that learning cannot be learned (at least not if we mean the same thing each time we mention 'learning') shows only that here, too, we need a better definition of innateness for the question of whether learning is innate to make sense.

The difficulties that arise when we try to give a suitable definition have led some philosophers to recommend that the term 'innate' be eliminated from science altogether (e.g. Griffiths 2002; Mameli and Bateson 2006). To get a flavour of these difficulties, consider that if one argues that an innate trait is one literally present at birth, it then makes no sense to say that someone is innately tall. One might say that an innate trait is one that is genetically determined, but no trait is built by genes alone – all require input from the environment, too. One might attempt to resolve this problem by defining an innate trait as one that is genetically 'specified', or genetically 'encoded', but this only creates a new problem, to say which of the genome's effects are to count as 'specified' or 'encoded' ones. The most promising definition equates innateness with developmental robustness or 'canalization' (Ariew 1999). On this view, an innate trait is one whose development is immune to variation in environmental conditions – change the developmental environment somewhat and the trait still develops in the same way. Such a definition is not without its own difficulties – how, for example, are we to specify which range of environmental alterations we should consider when determining whether a trait is innate? – but it seems to be the best of a bad lot.

Armed with this new definition of innateness, we can revisit the Santa Barbara School's claim that modules are innate. The philosopher David Buller casts doubt on this claim by drawing on the fact that human brains are highly 'plastic' (Buller 2005: 141). Our neural structure is subject to environmentally-induced change not just during infancy and childhood, but later in life, too. A widely-cited study conducted on a group of London taxi drivers suggested that their brains changed measurably as they went through 'The Knowledge' – the training period during which they learn how to navigate the city's streets (Maguire *et al.* 2000). Buller mentions a number of studies which suggest that specialised neural circuits can be built by interactions between the environment and plastic neural structure. The formation of these circuits is sensitive to variation in the developmental environment – in different environments, different circuits develop. If modules are specialised neural circuits, built by the action of the environment on plastic neural structure, then modules are not innate.

4. EVOLUTIONARY EPISTEMOLOGY: JAMES AND POPPER

Let us move on to our second topic, that of evolutionary episte-mology. One of the earliest applications of natural selection to the growth of scientific knowledge comes from the American philosopher William James. In the same 1880 essay that we discussed in chapter two, James notes that:

> A remarkable parallel, which to my mind has never been noticed, obtains between the facts of social evolution and the mental growth of the race, on the one hand, and of zoological evolution, as expounded by Mr Darwin, on the other.
>
> (James 1880: 441)

The primary purpose of James's essay is to use what he regards as a proper understanding of Darwinian explanation to undermine the 'so-called evolutionary philosophy of Mr Herbert Spencer' (ibid.: 422). James takes it that a good Darwinian will see social change in human societies as a form of evolution in its own right, in which individual people throw up different ideas, which go on to enjoy success or failure as a result of their fit with the social environment:

> Social evolution is a resultant of the interaction of two wholly dis-
> tinct factors: the individual, deriving his peculiar gifts from the
> play of physiological and infra-social forces, but bearing all the
> power of initiative and origination in his own hands; and second,
> the social environment, with its power of adopting or rejecting both
> him and his gifts. Both factors are essential to change. The
> community stagnates without the impulse of the individual. The impulse
> dies away without the sympathy of the community.
>
> (Ibid.: 448)

James sees individual genius as something inexplicable, but which nonetheless contributes to the course of social evolution by deter-mining the available variation for social selection to act upon: 'The causes of production of great men lie in a sphere wholly inaccessible to the social philosopher. He must simply accept geniuses as data, just as Darwin accepts his spontaneous variations' (ibid.: 445). Darwin, remember, thinks that the causes of varia-tion are beyond our ken. The environment determines which variations will survive and which will perish. James sees the indi-vidual genius as another inexplicable source of variation, whose ideas may or may not be accepted by the social environment.

James is arguing that a properly Darwinian view of social evolution keeps a central place for the individual genius. In this respect he opposes Spencer's scepticism regarding the importance of 'great men' in the history of science. James also hints that a Darwinian explanation of genius itself might be available. The mind, James says, is subject to various disturbances – ideas arise largely at random, so that 'according to the idiosyncrasy of the

individual, the scintillations will have one character or another.'
(ibid.) Scientific discovery then results from a kind of internal
selection process, by which random ideas are generated, and then
selected not by society, but by the individual's experience of the
world:

> The genius of discovery depends altogether on the number of these
> random notions and guesses which visit the investigator's mind. To
> be fertile in hypothesis is the first requisite, and to be willing to throw
> them away the moment experience contradicts them is the next.
>
> (Ibid.: 456–57)

James's views are in many respects close to those that the philosopher of science Karl Popper would articulate fifty years later in The
Logic of Scientific Discovery (Popper 1935). Popper, like James, sees the
growth of science as the result of interactions between a source of
novel proposals and a filter that weeds out some proposals and
retains others. More specifically, Popper says that science is a
process of 'conjecture and refutation' (ibid.) – the scientist
proposes bold hypotheses or conjectures, which are either refuted
or retained depending on how they mesh with the tribunal of
experience. Popper follows both Darwin and James in having no
embarrassment about confessing his ignorance – indeed Popper
confesses a lack of interest – regarding the causes of variation. In
the context of science, the causes of variation are the psychological
processes that explain the diverse conjectures put forward by
scientists.

Analogies always work both ways. Popper sees the growth of
knowledge as a form of conceptual evolution. He also sees organic
evolution by natural selection, even when it goes on in rudimentary organisms, as a form of trial and error learning:

> The method of trial and error is not, of course, simply identical
> with the scientific or critical approach—with the method of con-
> jecture and refutation. The method of trial and error is applied not
> only by Einstein but, in a more dogmatic fashion, by the amoeba also.
>
> (Popper 1962: 68)

The amoeba hypothesises that some blob of stuff is nutritious; if it is right, then the amoeba survives, and the hypothesis, too. If it is wrong, then the amoeba dies, and the hypothesis dies with it. Einstein represents an improvement over an amoeba, because Einstein's knowledge is gained in a similar way, but in a manner that obviates the need for all this death:

> The critical attitude might be described as the result of a conscious attempt to make our theories, our conjectures, suffer in our stead in the struggle for the survival of the fittest. It gives us a chance to survive the elimination of an inadequate hypothesis—when a more dogmatic attitude would eliminate it by eliminating us . . . We thus obtain the fittest theory within our reach by elimination of those which are less fit.
>
> (Ibid.: 68–69)

Popper then adds, in parentheses, a remark that is not defended but which will cause him significant trouble: '(By "fitness" I do not mean merely "usefulness" but truth. . . .)'.

Let us summarise Popper's view: he claims that science moves closer to the truth in the same way that a species adapts progressively to its environment. Hypotheses are compared with observational data. The hypothesis 'dies' if one of its predictions is contradicted by the data. If a hypothesis is compatible with the data, then it survives. But why suppose that compatibility with the data is indicative of truth?

Popper's argument from evolution to truth rests in part on the following seductive, yet misleading, connotations of the word 'fit'. If a theory is fit, then it fits the world. If it fits the world, the structure of the theory corresponds with the structure of the world. And if theory and world correspond, then the theory is true. However, the most selection ensures is adaptation with respect to the problem at hand – the 'fitness' of a theory is conformity with whatever data we have been able to gather. False theories, as well as true ones, often yield successful predictions. Therefore, a theory can conform with a body of data without conforming with the world.

What is missing from Popper's evolutionary epistemology is an argument that links the fitness of a theory to its truth. Popper's broader philosophical views offer scant hope of helping us formulate such an argument. He claims that when a conjecture is at odds with our data, we can reject it. Under these circumstances the conjecture has been 'falsified'. But he denies that we have good reason to believe a conjecture that has not been falsified is true. Indeed, he rejects any extrapolation from a theory's past successes in avoiding falsification to its prospects for future success against the tribunal of experience. Popper's picture of science is made more complex by his further claim that statements of the data, as well as theoretical hypotheses, have the status of conjectures. Since all conjectures are (for Popper) wholly tentative, if a theory avoids falsification this means only that one conjecture – the hypothesis – is consistent with another set of tentative conjectures – the data. For Popper, then, science is a process by which one set of tentative conjectures becomes adapted to another set of tentative conjectures. He offers no convincing arguments for why such a state of adaptedness has any bearing on the truth of either set of statements.

5. MEMES

In recent years, theorists from different disciplines have proposed evolutionary models of the sciences which take the loose analogy Popper draws between science and natural selection, and bring it into far closer formal alignment with the principles of modern evolutionary biology. Such models have their roots in works by the psychologist D. T. Campbell, who regards the growth of knowledge as a process of what he calls 'blind-variation-with-selective-retention' (Campbell 1974). They also owe a lot to Richard Dawkins' speculative remarks about the possibility of non-genetic evolution at the end of The Selfish Gene (1976), and to the philosopher of biology David Hull's pioneering studies of scientific change (Hull 1988).

In order to examine these more formal models of scientific evolution, we need to say a little about Hull's distinction between *replicators* and *interactors* (ibid.: 408). This is closely related to Dawkins' distinction between *replicators* and *vehicles*. When we introduced natural selection back in chapter two, we noted the widely acknowledged definition of selection as a process operating on entities that vary in their fitness, and which reproduce in such a way that offspring resemble parents. These conditions are stated in an abstract way, allowing that any set of entities, whether they are organisms, computer viruses, ideas or artworks, might be said to undergo selection, just so long as they reproduce, and offspring resemble parents. The virus in my computer is the 'offspring' of the virus in the computer that was the source of the infection; the scepticism I have about the existence of God is the 'offspring' of the atheism of David Hume, whose *Dialogues Concerning Natural Religion* converted me; the picture I painted is the 'offspring' of the Monet from which it was copied.

Hull claims that all selection processes – standard organic evolution included – require entities that play two distinct roles, roughly corresponding to the twin requirements that offspring resemble parents, and that parents differ in their fecundity. *Replicators* are entities that copy themselves, thereby ensuring trans-generational resemblance. Genes are usually thought of as replicators in organic evolution: offspring resemble parents, so it is said, because genes have the ability to make copies of themselves. *Interactors* are entities which cause replicators to appear in different proportions in the offspring generation, in virtue of their interactions with the environment. Fast-running wolves catch deer more efficiently, and as a result of this their genes are copied in greater proportions than the genes of slow-running wolves. In this particular case, wolves are interactors, while wolf genes are replicators; Hull's view allows that under some circumstances, a single type of entity (an asexually reproducing bacterium, for example) might act as replicator and interactor at the same time.

The replicator/interactor distinction raises many interesting questions that will not be addressed here. How, precisely, should we define these terms (Griffiths and Gray 1994)? Are genes the only replicators in organic evolution (Sterelny *et al.* 1996)? Could selection occur without replicators (Godfrey-Smith 2000)? If not, how can natural selection explain the initial emergence of replicators, which are clearly complex entities in their own right? Rather than looking at these questions in detail, let us instead look at Hull's application of the replicator/interactor distinction to scientific evolution. He claims that stability over time in science – whether that is stability of the theories believed by successive generations of scientists, or of styles of production of scientific documents, or of techniques used in the lab – should be explained by citing the transmission of reliably copied replicators. But the replicators in question are not genes: 'The mode of transmission in science is not genetic but cultural, most crucially linguistic. The things whose changes in relative frequency constitute conceptual change in science as elsewhere are "memes", not genes . . . ' (Hull 2001: 98).

Memes are cultural replicators. Richard Dawkins, who coined the term, gives us a list of exemplary memes which includes: 'tunes, ideas, catch-phrases, clothes fashions, ways of making pots or of building arches' (Dawkins 1976: 206). Let us focus for the moment on tunes. These are indeed particularly 'catchy' bits of culture, which often hop from person to person. A friend hums Singin' in the Rain one morning. I hear the tune, and find myself whistling it later on. By the evening, five or six of my colleagues have the same tune in their heads and at their lips. Thus the replicator makes copies of itself, and various interactions between noises and human brains cause this spread. The tune of Singin' in the Rain, Dawkins says, is a meme. Scientific theories, Hull says, are memes, too. They also make copies of themselves, hopping from brain to brain. Their rates of spread depend on the effects of diverse interactors – books, articles, conversations, tools – on the scientific environment. The criteria, conscious or unconscious, which scientists themselves bring to bear when assessing the merits of a theory constitute an important set of features of the selective environment of memes.

The meme concept has attracted a fair amount of hostile commentary. Even supporters of the evolutionary view of culture often argue that formalising cultural evolution using memes is misguided. Let me briefly review three fairly frequent complaints directed at the meme/gene analogy, before going on to pinpoint what is the most serious problem affecting memetics.

'Genetic units are discrete particles; culture is not composed of discrete units'

Critics sometimes argue that it makes no sense to think of an idea as a gene-like unit, which can be analysed in isolation. Ideas come packaged as interconnected systems – the idea of a god, for example, can only be understood when one also understands other ideas to which it is related. Depending on which religion we are discussing these might be ideas relating to paternity, maternity, grace, knowledge, love, vengefulness, and so forth. This also means that one cannot treat all instances of belief in a god as instances of the same type of meme. If belief in God only makes sense in the context of the system it features in, then one religion's 'belief in God' is only superficially similar to that of another. The anthropologist Adam Kuper summarises: 'Unlike genes, cultural traits are not particulate. An idea about God cannot be separated from other ideas with which it is indissolubly linked in a particular religion' (Kuper 2000: 180).

The memeticist is likely to accept that ideas need to be understood in context, and that not every belief in god is the same type of belief. But she is likely to add that genes are just like ideas in these respects. She might point out that genes depend for their effects on interactions with other genes in the organism. She might add that superficially similar genes, identified by their DNA sequences, can have very different roles in different species, so that it makes little sense to think of them as instances of the same type for the purpose of evolutionary analysis. Genes do not have a life of their own, in isolation from their specific web of relations to other genes, any more than ideas do. Even so, once we have specified some particular species, perhaps even some particular population, one can isolate the role of a gene in that context. The same, she might say, goes for memes. Ideas can be assigned individual roles once we specify a particular social context for investigation.

'Genetic units makes copies of themselves; cultural units do not'

Memes are supposed to be replicators. They are supposed to make copies of themselves. Now it is certainly true that ideas spread through groups of people. But it is less clear whether they do so by making copies of themselves. The anthropologist Dan Sperber complains that:

> . . . most cultural items are 're-produced' in the sense that they are produced again and again—with, of course, a causal link between all these productions—but are not reproduced in the sense of being copied from one another . . . Hence they are not memes, even when they are close 'copies' of one another (in a loose sense of 'copy', of course).
>
> (Sperber 2000: 164–65)

Recall the example of my whistling *Singin' in the Rain*. I whistle it because I heard the tune earlier in the morning. In a sense, a reproduction has been made of the tune. But although my performance resembles the earlier one, is mine copied from the earlier one? Perhaps it is: perhaps I listen very carefully to the tune, take efforts to memorise it, and whistle it myself. But probably I do not do this. More likely I hear a little of the tune and think 'Aha! That chap is humming *Singin in the Rain*! Such a fine tune!' The tune is already familiar to me, I have no need to listen carefully, and I begin to whistle it myself. In this second case it is somewhat strained to say that my version of the tune is a *copy* of the one I hear. It seems more appropriate to say that my hearing the tune *triggers* the performance of a tune that is already in my repertoire. Sperber understands replication as copying in the strong sense, rather than as the triggering of a resembling performance. He goes on to argue that most cultural reproduction is of the triggering type, not the copying type. As a result replication is comparatively rare, and there are fewer instance of meme-like reproduction than at first meet the eye. Spelling out in precise terms what the difference is between copying and triggering may be difficult, but Sperber is right to remind us that there are many different ways for the same idea, or the same behaviour, or the same tune, to be 'reproduced' through a population. What is more, if the successes of organic evolutionary theory are anything to go by, evolutionary theory becomes enlightening when we are able to characterise in some detail the modes that reproduction takes. This is why Mendel's genetic laws are important: Mendel's laws tell us something about the general patterns of parent/offspring relations, which in turn help us to explain the makeup of successive generations of a population. Our theories of cultural evolution, if they are to be enlightening, need to do more than assert that culture contains varied ideas which are reproduced at different rates. In this sense, the mere claim that culture evolves is not sufficient to make cultural evolutionary theory informative. These theories also need a rich enough vocabulary to capture different modes of cultural reproduction, and they need to investigate how those modes of cultural reproduction affect the composition of successive cultural generations.

'Genes form lineages; cultural units do not'

A worry that is closely related to Sperber's draws on the fact that ideas do indeed spread through populations when individuals learn from each other, but these ideas do not always form *lineages* in the ways that genes do (Boyd and Richerson 2000). In principle, I could look into my genome and say (for most of my genes, at least), which came from my father and which from my mother. Each gene is derived from a single individual, in such a way that we might trace a lineage back through time. Can we do this for cultural items? Not always. Consider my knowledge of the tune of *Singin' in the Rain*. It is unlikely that there is a single source from which this knowledge is derived. It is unlikely, for example, that I learned this tune because one other person whistled it. I probably picked the tune up over time, from exposure to parents, friends, various showings of films, and so forth. The knowledge of a tune like *Singin' in the Rain* spreads through a population, and various facts might make it spread more quickly than other tunes. Yet it is misleading to call *Singin' in the Rain* a meme, because unlike genes, people who know the tune of *Singin' in the Rain* have rarely inherited it from a single individual. What facts might make one tune more likely to spread than others? In part, of course, we can point to facts that make an individual who knows the tune more likely to whistle it, and to facts that make an individual who hears it more likely to remember it. But a tune could score comparatively poorly on these characteristics and still spread faster than its competitors simply because it is ubiquitous. If a record company ensures that a melody is played through all available radio and TV networks, then even a tune that is comparatively un-catchy will quickly become known by millions. This underlines an important limitation for memetics. In organic evolution, the swift spread of some variation through a population typically indicates that the variation in question confers high reproductive success on its bearers. Things are more complicated at the cultural level. We cannot infer from the swift spread of a tune through a population that the tune has features than make it likely to hop from mind to mind. The tune may not be especially 'contagious' or 'catchy' at all; the tune's producers may just be powerful enough to make it ubiquitous, hence more likely to be learned than far catchier, but more poorly-funded, competitors. Once again, it is important that our cultural evolutionary theories are rich enough to document the diverse reasons why an idea may spread, and the memetic theory, by drawing a very close analogy between organic and cultural evolution, threatens to obscure the important distinction between contagious and power-assisted spread.

6. CULTURAL EVOLUTION WITHOUT MEMES

Martin Gardner (quoted in Aunger 2000: 2) complains that: 'memetics is no more than a cumbersome terminology for saying what everybody knows and that can be more usefully said in the dull terminology of information transfer'. Even if memeticists are right that culture evolves, this is not informative unless they go some way to spelling out the details of how culture evolves. The memeticist might answer with a case-study. Suppose we are trying to explain why more people buy new Minis than buy new Beetles. We could do so by suggesting that one meme – the inclination to buy a new Mini – is fitter than another – the inclination to buy a new Beetle. What makes one fitter than the other? Perhaps Minis look cooler than Beetles, or perhaps they run better, or they are cheaper. The result is that more copies are made of the Mini-purchasing meme than the Beetle-purchasing meme, and these memes cause differential purchasing behaviour on the part of their bearers. If case-studies like this are the best we can come up with, Gardner's objection is reinforced. This memetic explanation is merely a cosmetic repackaging of the kind of story an economist, or a psychologist, might tell about why more people buy Minis than Beetles.

The assertion that cultural change can be understood in terms of various factors that explain the relative successes and failures of different ideas verges on the trivial. A useful theory of cultural evolution needs to offer some insight regarding what factors need to be taken into account in explaining the changing composition of a population of ideas, and these factors need to be unlikely to be noticed by students of more traditional disciplines like economics, or psychology. Memetics is particularly unlikely to yield an informative cultural evolutionary theory of this kind. Its proponents appear to think that because genetic models of evolution have been largely successful in the organic realm, similar models must be the best ones to use for the cultural realm. Genes are understood as discrete particles that are faithfully copied; consequently memes are understood as discrete particles that are faithfully copied. Perhaps memeticists think that the pioneers of the modern theory of evolution – people like R. A. Fisher – showed that natural selection can only work when inheritance is 'particulate'. Remember the problem we examined in chapter two, which Fleeming Jenkin raised for natural selection. If offspring are always intermediate in character between their parents, then, Jenkin said, it seems that beneficial mutations will not be added up and preserved by selection, but instead they will be washed away over time by the action of 'blending inheritance'. Fisher did not solve this problem by arguing that natural selection can work only if inheritance is underpinned by the transmission of discrete particles (as opposed to a general blending of parental traits). Instead, Fisher argued that natural selection would only be able to produce cumulative adaptation within a system of 'blending' inheritance if mutation rates were very high – certainly higher than observed genetic mutation rates. Fisher's achievement was to show how a system of particulate inheritance would enable natural selection to operate effectively even with low genetic mutation rates.

Fisher's work immediately prompts a series of questions we can ask about cultural evolution. What might we mean by the 'rate of cultural mutation'? How could we measure it? Is the cultural mutation rate higher than the genetic mutation rate? Is it high enough for natural selection to be able to operate effectively without the faithful replication of cultural 'particles'? Is each cultural trait of an individual organism in fact a blend, assembled from diverse influences (such as parents, siblings and authority figures)? How does cultural inheritance affect the natural selection of organisms? How, for example, does cultural inheritance – which can perhaps maintain the presence over time of traits in social groups – affect the operation of natural selection at the level of the group? All of these questions are best answered using a combination of rigorous statistical modelling coupled to detailed empirical investigation – just the techniques that enabled the pioneers of the modern synthesis to make natural selection a well-understood and well-confirmed explanation for organic evolution. The memetic view has a tendency to obscure the importance of questions like these. That is why the most constructive work in cultural evolutionary theory has been done by those who are sceptical of memes – people like evolutionary anthropologists Robert Boyd and Peter Richerson, who try to answer just these questions (Richerson and Boyd 2005).

To give a hint of the promise that Boyd and Richerson's approach holds for the understanding of culture, consider their discussion of technological innovation (ibid.: 52–54). They begin by telling the story of the development of the modern ship's compass. It is a complex one, which starts with the discovery that naturally-occurring magnetite has a tendency to directional orientation.

Further refinements are spread over centuries and continents. They include the production of magnetite needles that can be floated in a bowl of water, the mounting of a magnetic needle on a vertical pin bearing, the addition of iron balls that cancel out the distorting effects of steel-hulled ships, and the perfection of various systems that damp the response of the compass to the ship's rocking.

This account, while recognisably evolutionary in its commitment to a Darwinian form of gradualism, is unlikely to startle any historian of technology. It is no surprise to learn that innovation often proceeds through the accumulation of many small steps, which have taken place over a considerable breadth of time and space. Things get more interesting when Boyd and Richerson begin to ask comparative questions. European empire-builders successfully invaded the Americas; the Americans did not invade Europe. Why did it happen this way round? They follow Jared Diamond (1996) in attributing it to the greater pace of technological innovation in European, compared with American, societies. In chapter two we noted Darwin's own recognition – a purely statistical insight – that the size of a social group can affect the chances of useful technological inventions being produced in it. Boyd and Richerson offer a similar form of explanation for the difference in innovative pace in the two landmasses, which draws once again on Diamond's work. They consider innovation a function of the likely rate at which cultural mutation can be generated, and of the chances of advantageous cultural mutations increasing in frequency once produced. They suggest that:

> . . . the greater size of the Eurasian continent, coupled with its east-west orientation, meant that it had more total innovation per unit of time than smaller land masses, and that these innovations could easily spread through the long east-west bands of ecologically similar territory.
>
> (Richerson and Boyd 2005: 54)

Suggestions like this are certainly speculative, but they show the potential for explanatory novelty offered by the evolutionary approach to culture.

SUMMARY

Darwin's work has influenced epistemology in two main ways. The first is direct. Prior to Darwin, philosophers had long been divided on the existence of innate knowledge. On the one hand, it seemed to many that we know things that we have not had to learn. But if learning does not account for the possession of a true belief, what does? Darwin's evolutionary theory immediately suggests a plausible mechanism that might explain this – namely, natural selection – and his work thereby offers the promise of rendering innate knowledge respectable. Darwin's work has also influenced epistemology in an indirect fashion. Natural selection can be stated in an abstract way, which allows us to see entities other than organisms as subject to selection processes. Entities of any type can be said to evolve by natural selection, so long as they vary, they reproduce and offspring resemble parents. So-called 'evolutionary epistemology' claims that scientific theories meet these conditions, and it consequently studies scientific change as an evolutionary process. Recently, evolutionary epistemologists have made widespread use of the meme concept, regarding scientific theories, and ideas in general, as memes. Memes are supposed to be cultural analogues of genes. They are replicators – that is, entities which make copies of themselves – and they underlie cultural inheritance. We saw in this chapter that there are reasons to be sceptical of the meme/gene analogy. It is far from clear that all ideas are replicators (although some might be), and it also unlikely that ideas always form lineages (although sometimes they do). More importantly, even if memetics' defenders are right to say that culture evolves, and that cultural evolution consists in the differential spread of different types of meme, it is unclear how much insight this brings that could not be had just as well by using models from psychology or economics. This is not to say that no cultural evolutionary theory has value, but such theories need to examine how ideas are reproduced, how they mutate, how the structure of a population of ideas affects the prospects of that population, and so forth. These are the kinds of questions that needed to be answered before Darwin's theory of natural selection could be applied in a detailed manner to organic evolution, and the same questions need to be asked in the cultural realm.

FURTHER READING

Elliott Sober has a clear discussion of the issues surrounding the relative evolutionary merits of belief-forming mechanisms liable to error and those that are more accurate, as well as the merits of innate belief compared with learning:

Sober, E. (1994c) 'The Adaptive Advantages of Learning and A Priori Prejudice', in E. Sober, *From a Biological Point of View*, Cambridge: Cambridge University Press.

A good place to turn to for Popper's evolutionary epistemology is:

Popper, K. (1962) 'Conjectures and Refutations' in K. Popper, *Conjectures and Refutations*, London: Routledge.

An extended statement of Campbell's evolutionary epistemology appears in:

Campbell, D. T. (1974) 'Evolutionary Epistemology', in P. Schilpp (ed.) *The Philosophy of Karl Popper*, LaSalle, IL: Open Court.

A useful collection of essays on memes – some hostile, some friendly – was published a few years ago:

Aunger, R. (2000) *Darwinizing Culture*, Oxford: Oxford University Press.

Daniel Dennett also discusses the meme concept in a balanced and characteristically lively fashion towards the end of his book on Darwinism:

Dennett, D. (1995) *Darwin's Dangerous Idea: Evolution and the Meanings of Life*, London: Allen Lane.

A particularly sophisticated overview of modern cultural evolutionary theory can be found in:

Richerson, P. and Boyd, R. (2005) *Not by Genes Alone: How Culture Transformed Human Evolution*, Chicago: University of Chicago Press.

Eight

Politics

I. DARWIN AND THE RIGHT

As befits a Victorian, Darwin's political views do not fit easily into the categories we use today. Take the example of race. As we have seen, Darwin believed that all human races were very closely related to each other, and he supported his case by arguing that only the finest of gradations could be observed between them. He was passionately opposed to slavery. He dismissed exceptionless generalisations about the makeup of different races. But he was no racial egalitarian. He believed that there were good rules of thumb about racial psychology: 'Their mental characteristics are likewise very distinct ... Every one who has had the opportunity of comparison, must have been struck with the contrast between the taciturn, even morose, aborigines of S. America and the light-hearted, talkative negroes' (*Descent*: 198).

More to the point, Darwin was happy to describe races in terms of higher and lower, and he had no qualms about likening the lower human races to the higher apes. He believed that whites were the highest of all races. Such descriptions can be observed, for example, in Darwin's evolutionary explanation for why the current gap in advancement between man and his nearest living relative will in time become even greater:

> At some future period, not very distant as measured by centuries, the civilised races of man will almost certainly exterminate, and replace, the savage races throughout the world. At the same time the anthropomorphous apes, as Professor Schaaffhausen has

remarked, will no doubt be exterminated. The break between man and his nearest allies will then be wider, for it will intervene between man in a more civilised state, as we may hope, even than the Caucasian, and some ape as low as a baboon, instead of now between the negro or Australian and the gorilla.

(Ibid.: 183–84)

As we will see, just as it is hard to pigeon-hole Darwin's views on race, it is also difficult to pin down the relationship between Darwin's theory and political thought in general. It is perhaps tempting to associate evolution by natural selection with the hard right. Darwin's theory has sometimes been taken to entail the grim view that social improvement is best achieved by struggle between individuals, in which the weak must be allowed to perish if communities are to thrive. Darwin was aware of this interpretation of his theory, and distanced himself from it. Only a few weeks after the *Origin* was published he wrote in a letter to Lyell: 'I have received, in a Manchester newspaper, rather a good squib, showing that I have proved "might is right," and therefore that Napoleon is right, and every cheating tradesman is also right' (Darwin 1905: 56–57).

Darwin is also sometimes taken to bolster a rather different strand of right-wing thinking, by undermining the optimism more usually associated with the left regarding the ability of social reform to eliminate inequalities. If Darwin is right, the argument goes, then some inequalities (perhaps gender inequalities) owe themselves directly to deep facts about human nature, which no amount of social reform can alter. The goal of this chapter is to assess the degree to which Darwin's ideas support any one political stance, right or left.

2. DEGENERATING SOCIETY

Darwin believed that the contorted form of selection to which our species is subjected presented 'a most important obstacle in civilised countries to an increase in the number of men of a superior class . . . ' (*Descent*: 163). He worried that the position of the virtuous was precarious, for those prone to vice would tend to out-compete them, thereby lowering the elevation of civilisation. He elaborates the problem, citing William Greg (a mill-owner, who had been at Edinburgh University at the same time as Darwin) with great approval:

> Thus the reckless, degraded and often vicious members of society, tend to increase at a quicker rate than the provident and generally virtuous members. Or as Mr Greg puts the case: 'The careless, squalid, unaspiring Irishman multiplies like rabbits: the frugal, foreseeing, self-respecting, ambitious Scot, stern in his morality, spiritual in his faith, sagacious and disciplined in his intelligence, passes his best years in struggle and in celibacy, marries late, and leaves few behind him. Given a land originally populated by a thousand Saxons and a thousand Celts—and in a dozen generations five-sixths of the population would be Celts, but five-sixths of the property, of the power, of the intellect, would belong to the one-sixth of Saxons that remained. In the eternal "Struggle for Existence", it would be the inferior and less favoured race that had prevailed—and prevailed by virtue not of its good qualities but of its faults.'
>
> (Ibid.: 164)

A high standard of morals, in other words, is a fragile feature of today's human populations. How is this to be reconciled with Darwin's earlier evolutionary account (which we covered in chapter six) of the emergence of the moral sense, in which natural selection favours virtue? This account, he says, was restricted to an explanation of moral progress in 'the advancement of man from a semi-human condition to that of the modern savage'. Natural selection no longer has the same character in civilised societies, with the result that the reversal of this progress is an immediate danger (ibid.: 158). This downward trend is most visible when we look to the maintenance of mental and physical health:

> With savages, the weak in body or mind are soon eliminated; and those that survive commonly exhibit a vigorous state of health. We civilised men, on the other hand, do our utmost to check

the process of elimination; we build asylums for the imbecile, the maimed, and the sick; we institute poor laws; and our medical men exert their utmost skill to save the life of every one to the last moment . . . Thus the weak members of civilised societies propagate their kind. No one who has attended to the breeding of domestic animals will doubt that this must be highly injurious to the race of man. It is surprising how soon a want of care, or care wrongly directed, leads to the degeneration of a domestic race; but excepting in the case of man himself, hardly anyone is so ignorant as to allow his worst animals to breed.

(Ibid.: 159)

In a moment, I will evaluate Darwin's response to these fears for the degeneration of man. First, we should note that these passages bring out another difference between Darwin's conception of selection and our own. Darwin appears to believe that natural selection is either malfunctioning in civilised humans, or not operating at all. On the modern view, fitness is tied directly to reproductive success. If Greg is right that the reproductive rate of the Irish is higher than the reproductive rate of the Scots, with the result that the Irish will soon overrun the Scots, then this is to say that the Irish are fitter than the Scots, which, in turn, is to say that selection favours the Irish. Whether a world full of Irish is preferable to a world full of Scots is something that we may have opinions on, but those opinions are not dictated by evolutionary considerations. It therefore makes no sense on the modern conception to say that selection is *malfunctioning*, for selection favours the reproductively fitter variant, regardless of how we might evaluate the virtue of that fitter form.

Darwin, by contrast, constantly stresses in his presentations of natural selection his belief that 'natural selection works solely by and for the good of each being', and there is evidence that he intends this 'good' to be read in a moral sense. Consider, in support of this conjecture, Darwin's summary of the argument in favour of natural selection. We looked at this passage in detail back in chapter two, but it is important to note that where modern conceptions of selection might recognise only one argumentative step, Darwin recognises two. The crucial sentences are these:

> I think it would be a most extraordinary fact if no variation ever had occurred useful to each being's own welfare, in the same way as so many variations have occurred useful to man. But if variations useful to any organic being do occur, assuredly individuals thus characterised will have the best chance of being preserved in the struggle for life; and from the strong principle of inheritance they will tend to produce offspring similarly characterised.
>
> (*Origin*: 169–70)

Today we would content ourselves with noting that if variations occur that promote an organism's ability to survive and reproduce, then given the principle of inheritance, such variations will appear in offspring, too. But Darwin adds an intermediate step that is, from the modern perspective, a distraction. First, he says, we should expect some variations to occur that are useful to each being's own *welfare*, and second, variations of this kind will tend to give organisms the best chance of being preserved in the struggle for existence. This conceptual linkage between natural selection and that which promotes individual welfare enables Darwin, and Greg, to hint that natural selection is malfunctioning when the variations which are preserved are (as in the case of the sick) injurious to welfare. This moralised reading of natural selection enables us to understand Darwin's bleak optimism regarding the end result of the struggle for existence in nature:

> When we reflect on this struggle, we may console ourselves with the full belief, that the war of nature is not incessant, that no fear is felt, that death is generally prompt, and that the vigorous, the healthy, and the happy survive and multiply.
>
> (Ibid.: 129)

3. SOCIAL DARWINISM

To recap, Darwin is concerned that our refusal to allow the unhealthy (whether in body or mind) to perish is likely to lead to a degeneration of the species. He does not jump to the conclusion that we should allow weaker members of society to die. He believes that the support we give to the ill and sick has adverse consequences for the general good of the species, but he denies that these consequences justify withdrawal of that support. Darwin argues that it is the 'noblest part of our nature' that prompts our sympathy with the helpless, and if we were to look coldly on their suffering it would lead to a deterioration of that noble nature. The most obvious cure for social degeneration, then, would be worse than the disease. Instead, we should hope that some of the ills that arise from the assistance given to the weak will be mitigated by their not having offspring, either because the weak will have diffi-culty in marrying, or because they will choose not to do so:

> The surgeon may harden himself while performing an operation, for he knows that he is acting for the good of his patient; but if we were intentionally to neglect the weak and helpless, it could only be for a contingent benefit, with an overwhelming present evil. We must therefore bear the undoubtedly bad effects of the weak surviving and propagating their kind; but there appears to be at least one check in steady action, namely that the weaker and inferior members of society do not marry so freely as the sound; and this check might be indefinitely increased by the weak in body or mind refraining from marriage, though this is more to be hoped for than expected.
>
> (*Descent*: 159–60)

One cannot sever all links between Darwin and the later eugenicists of the nineteenth and twentieth centuries. Darwin appears to share the eugenicists' readiness both to acknowledge a threat of social decline, and to blame that decline on alleged hereditary deficien-cies among individuals. All were concerned that uncontrolled human breeding, especially in a context of state-provided care that could prolong the lives of the needy, might lead eventually to a decline in the vigour of human communities. But Darwin refuses to draw the inference that this justifies heavy-handed interven-tions to actively limit human reproduction. By pinning his hopes on voluntary modes of birth control and the natural regulation of reproduction, Darwin's views remain distant from those of the twentieth century's most notorious eugenic criminals.

Darwin also distances himself from the strongest forms of Social Darwinism, which argue that the struggle for life (either between individuals or between social groups) is a justified and effective means to human progress. He does, of course, believe that natural selection in the past – especially the struggle between groups, which promotes sympathy and lays the foundations of the moral sense – has produced many valuable human traits. But it does not follow from this that we should continue to act to promote the intensity of social struggle now. Darwin argues that 'highly civilised nations . . . do not supplant and exterminate one another as do savage tribes' (ibid.: 168–69). We are surely meant to read this as a moral argument: Darwin is not claiming that European nations, for example, never exterminate one another; rather, he is making his view known that true civility rules out such behaviour. But Darwin adds that once natural selection has put in place such traits as sympathy and intelligence, there is no reason to suppose that selective struggle will continue to provide the most effective means of ensuring progress:

> The more efficient causes of progress seem to consist of a good education during youth whilst the brain is impressible, and of a high standard of excellence, inculcated by the ablest and best men, embodied in the laws, customs and traditions of the nation, and enforced by public opinion.
>
> (Ibid.: 169)

This strand of Darwin's own Social Darwinism, which rejects struggle as the proper means to progress, and which advocates education and strong role models as the primary means to improvement in civilised societies, is hard to reconcile with an image of Darwin as the political enemy of those reformers who pin their hopes for social improvement on the education system.

Finally, it is worth mentioning a modern argument against eugenics that Darwin did not subscribe to. Darwin denied the moral legitimacy of controlled breeding programmes, but he did not deny their efficacy. Today, one frequently reads that modern genetics – genetics that Darwin was in no position to know about – teaches us that controlled breeding programmes would never have been effective in eliminating hereditary diseases.

We should indeed be sceptical of the efficacy of eugenic programmes. Many hereditary diseases, we now know, cannot possibly be eliminated from a population by sterilising the affected members, or by ensuring they do not breed. This is because our genes come in pairs, and many inherited diseases are only had by people with two copies of the disease gene. Any population will contain many people who are perfectly healthy, and who nonetheless have one copy of the disease gene. When two such people reproduce, there will be a one-in-four chance that their offspring will have two copies of the disease gene, hence that they will have the disease. This means that the sterilisation of all disease sufferers would not be enough to eradicate many heredi-tary diseases: we would have to somehow also control the breeding of all those healthy individuals who carry the gene.

It is true, then, that effective controlled breeding programmes would be enormous undertakings, intruding into the lives of the healthy, as well as the diseased. Even so, one could recognise this fact and still endorse eugenics. Historians Diane Paul and Hamish Spencer have argued that the eugenicists of the early twentieth century were well aware of it. They made no simple factual error: 'By the 1920s, they well understood that the bulk of genes for mental defects would be hidden in apparently normal carriers. For most geneticists, this appeared a better reason to widen eugenic efforts than to abandon them' (Paul and Spencer 2001: 112).

Our modern opposition to eugenic programmes does not rest solely on the difficulty we estimate in building effective breeding programmes. It also rests in part, as Darwin's opposition rested, on the belief that an improvement in the overall health of a group of people does not justify in any straightforward way either the systematic neglect of, or the debasing interference with, the lives of individuals.

4. POLITICS AND HUMAN NATURE

It is rare now to see Darwinians argue in favour of harnessing the power of natural selection in order to improve the human species.

But that does not mean that Darwinian thinking is no longer believed to have political impact. These days, those who argue that Darwin's thinking has political relevance typically begin by pointing out the importance of a proper account of human nature for effective policy interventions. A good account of human psychology can make a difference to what policies we choose to implement. How, for example, might a government go about encouraging people to save for their retirements? The American lawyer and philosopher Cass Sunstein argues that schemes which give employees the opportunity to save a percentage of their salary will see greater uptake if saving is the default option, which employees must actively choose to opt out of, rather than the option which employees must actively choose to opt in to (Sunstein 2005). Sunstein bases this policy recommendation on psychological work demonstrating that people have a kind of inertia when it comes to making choices.

Some hold that Darwin should make a difference to policy-making because evolutionary psychology is particularly well placed to illuminate our psychological makeup. They tend to argue that the adaptive heuristic, which we investigated in chapter five, is an important tool here. Thornhill and Palmer frequently suggest in their book on the evolutionary psychology of rape that one of its primary values will be to enable policy makers to confront rape more effectively (Thornhill and Palmer 2000). The British philosopher Helena Cronin has argued that social policy would benefit from attending to work on the evolutionary psychology of sex differences:

> How could responsible social policy *not* be informed by an evolutionary understanding of sex differences? All policy-making should incorporate an understanding of human nature, and that means both female and male nature. Remember that if policy-makers want to change behavior, they have to change the environment appropriately. And what's appropriate can be very different for women and for men. Darwinian theory is crucial for pointing us to those differences.
>
> (Cronin 2004: 61)

Some have argued not merely that evolutionary psychology is a tool that will help us to understand our psychological makeup, but that evolutionary psychology reminds us that this makeup must be understood as a fixed core of human nature. They go on to argue that evolutionary psychology undermines a strand of left-wing thinking that regards claims about human nature with suspicion. If human nature were fixed, it would not follow that policy must always be impotent in the alteration of behaviour. But these Darwinians say that instead of trying to alter human nature itself, we should seek to manipulate environmental circumstances so that fixed human natures will yield the behaviours we are aiming at. To phrase things in the language of computers, this view has it that we can control the inputs people receive, and thereby the outputs they produce, but not the programs that match different inputs to different responses. Once again, Helena Cronin is a strong advocate of this view:

> Certainly, human nature is fixed. It's universal and unchanging, common to every baby that's born, down through the history of our species. But human behavior, which is generated by that nature, is endlessly variable and diverse. After all, fixed rules can give rise to an inexhaustible range of outcomes. Natural selection equipped us with the fixed rules—the rules that constitute our human nature. And it designed those rules to generate behavior that's sensitive to the environment. So, the answer to 'genetic determinism' is simple. If you want to change behavior, just change the environment. And to know which changes would be appropriate and effective, you have to know those Darwinian rules. You need only to understand human nature, not to change it.
>
> (Ibid.: 55)

5. DARWIN AND THE EQUALITY OF THE SEXES

No one should doubt that a knowledge of what makes us tick should make a difference to the policies we implement. We should also be prepared to accept that evolutionary reflection can be an important tool (along with direct observation and experiment) in uncovering our psychological makeup. However, we need to make sure that this work is done with customary scientific caution. There is a particularly seductive form of inference that begins with a piece of widely-believed, yet wholly prejudicial, folk wisdom – something that 'everybody knows' about how humans generally think or behave. One supplements this with a plausible evolutionary story about why it would have been advantageous to our ancestors to think or behave like this. This match between folk wisdom and evolutionary explanation is then taken to bolster the view that folk wisdom describes a genuine feature of our evolved nature. Darwin falls into this trap in his discussion of differences between human males and females in *Descent* (for a detailed examination of these passages see E. Richards 1983). He begins by listing physical differences: 'Man on an average is considerably taller, heavier, and stronger than woman, with squarer shoulders and more plainly-pronounced muscles' (*Descent*: 621). Soon, he moves onto differences in character: 'Man is more courageous, pugnacious and energetic than woman, and has a more inventive genius' (ibid.: 622). He waits until a few pages later before giving any evidence to back these assertions, and his evidence is flimsy. He notes, for example, that:

> I am aware that some writers doubt whether there is any such inherent difference [in the mental powers of man and woman]; but this is at least probable from the analogy of the lower animals which present other secondary sexual characteristics. No one disputes that the bull differs in disposition from the cow, the wild-boar from the sow, the stallion from the mare, and, as is well known to the keepers of menageries, the males of the larger apes from the females.
>
> (Ibid.: 629)

One might object. Analogy with other animals suggests we have some reason to expect differences between the human sexes, but this hardly confirms Darwin's specific assertion about the character of those differences. The remainder of Darwin's evidence draws on three sources: what 'is generally admitted' about sex differences, the historical record of celebrity in intellectual pursuits, and a plausible evolutionary hypothesis that is consistent with these largely anecdotal observations. Thus Darwin tells us (without citing any particular evidence) that:

> Man is the rival of other men; he delights in competition, and this leads to ambition which passes too easily into selfishness. These latter qualities seem to be his natural and unfortunate birthright. It is generally admitted that with woman the powers of intuition, or rapid perception, and perhaps of imitation, are more strongly marked than in man . . . .
>
> (Ibid.: 629)

He tells us of the high regard in which men are held:

> The chief distinction in the intellectual powers of the two sexes is shewn by man's attaining to a higher eminence, in whatever he takes up, than can woman—whether requiring deep thought, reason, or imagination, or merely the use of the senses and hands. If two lists were made of the most eminent men and women in poetry, painting, sculpture, music (inclusive both of composition and performance), history, science, and philosophy, with half-a-dozen names under each subject, the two lists would not bear comparison.
>
> (Ibid.: 629)

Finally, Darwin shows that these supposed inequalities (if we grant them) are explained by supposing that competition for women has driven men to greater and greater heights of genius (of which Darwin thinks patience, or 'unflinching, undaunted perseverance' is a primary constituent). Since sexual selection is prevalent throughout nature, and since it is capable of producing significant differences between the sexes in other species (such as the peacock, with a gaudy tail, and the peahen, with a plain one), the ability of sexual selection to account for differences in mental endowment between men and women should further convince us that such differences are real:

> Amongst the half-human progenitors of man, and amongst savages, there have been struggles between the males during many

generations for the possession of the females. But mere bodily
strength and size would do little for victory, unless associated
with courage, perseverance, and determined energy.

(Ibid.: 630)

What Darwin has given us here is a coherent and suggestive case.
But he certainly has not established the existence of the sex differences that sexual selection is invoked to explain. An alternative
coherent and suggestive case might say that men and women are
equally creative, and women have simply been discouraged from
expressing that creativity, or prevented from attaining celebrity in
spite of their creativity. This would account for the difficulties
Victorians had in drawing up lists of famous female contributors to
the arts and sciences, and it would also account for the general
acceptance among Victorians of women's creative inferiority. If
there is no true difference in the levels of creativity of men and
women, then sexual selection has nothing to account for in this
respect.

I have not argued that Darwin is mistaken about the character
of sex differences in humans; I have argued that the evidence he
gives to back his view is weak. In recent years others have tried to
supply more evidence to bolster it (e.g. Miller 2000). The point of
this section is simply to highlight the hazards of accepting a set of
claims about human nature based on a plausible evolutionary
story coupled to scant data about human psychology.

6. SEX DIFFERENCES TODAY

One might think that the excursion of the preceding section has
been primarily of historical interest. So what if Darwin jumped
hastily to conclusions about sex differences and their evolutionary
explanation? When today's evolutionary psychologists inform us
of the workings of the human mind, they attend closely to
evidence, they cautiously document patterns of similarity and
difference in human behaviour and human psychology, they
propose evolutionary hypotheses that make sense of those
patterns, and they test the assumptions of those hypotheses with

care. This is, indeed, what happens in the best evolutionary psycho-logical work. But the philosopher David Buller has raised doubts about some modern work on sex differences (Buller 2005).

Buller considers in detail the widely-accepted view that males and females have strikingly different mate preferences. According to the version of this hypothesis defended by David Buss, men prefer mates who are young, while women prefer mates with high levels of resources (Buss 1994). Buss argues that these sex differences are precisely the ones we should expect evolution to have endowed us with. His theoretical argument begins with the fundamental asymmetry that lies at the heart of male and female reproduction. This asymmetry is what drives asymmetrical mate preferences in all kinds of species. For females, reproduction is a process that demands huge investment – far greater than that demanded of males. Both males and females must attract a mate and have sex. But the physiological costs to a female in producing an egg are higher than the costs to a male in producing a sperm cell. Once fertilisation is completed, the male is effectively free to find more mates, and have more offspring. For the female, however, she must devote resources to gestation, and perhaps to lactation, too. Since females must expend greater resources than males if they are to reproduce successfully, it follows that females stand to lose far more than males if they make a poor choice of mate. Females, then, should evolve to be more discriminating regarding whom they choose to mate with; meanwhile, there should be considerable competition between males for opportuni-ties to mate with females (Buss 1999: 102–3).

Buss does not argue that evolutionary considerations predict that human males will not care at all about the attributes of their long-term mates. Males, as well as females, will seek healthy mates, because a healthy mate produces healthy children, and is likely to be up to the demands of parenthood. Buss adds that both sexes tend to favour mates who are intelligent, kind and under-standing (ibid.: 134). But in many other respects their mate preferences are driven in quite different directions by the different adaptive problems to which they have been exposed. While males seek mates who have strong reproductive potential, females seek mates who are able and willing to provide resources of care after birth. Since these qualities are hard to perceive directly, males and females have instead evolved to find mates attractive according to detectable features. Men place particular emphasis on features associated with youth, health, and consequently reproductive potential. More specifically, they look for 'full lips, clear skin, smooth skin, clear eyes, lustrous hair, good muscle tone and body fat distribution', as well as behavioural indicators such as 'bouncy youthful gait, an animated facial expression, and a high energy level' (ibid.: 139). Women place particular emphasis on features associated with the possession of resources; specifically they look for economic wealth, social standing, and more readily perceptible characteristics associated with these attributes, such as employment type, dress, educational achievement and age. Moreover, they focus on the qualities that lead a man to acquire and retain resources – qualities like ambition and dependability. Finally, they are attracted to men who are willing to invest their resources in children. Thus, evolutionary psychology predicts, explains and vindicates what everybody already knows and only a fool would deny about the differences between what men and women want.

My presentation of Buss's position has traded so far on the fit between a fairly abstract evolutionary argument and folk wisdom regarding differences in the mate preferences of males and females. What evidence is there that men really do prefer young women, and that women prefer men with high resources? Buller thinks the evidence is not as strong as one might think. Let us begin with female mate choice. Some scientists – evolutionists included – have argued that female preference for high-status males is a result of recent socioeconomic inequalities, rather than of selection for mate preferences in the distant past. They say that there is no ancient evolutionary preference for high-status males; rather, both males and females seek wealth and resources, and the recent economic climate has meant that for women the most effective way to do this is through selecting a high-status mate, rather than by directly seeking high status for themselves. Buss's response to this is to cite evidence suggesting that high-earning women typically have the strongest preference of all women for high-status men (ibid.: 124). Buller, however, casts doubt on whether women have a general preference for high-status males at all.

Buller criticises a 1990 study by Townsend and Levy, in which 112 women were asked to look at photographs of two male models. One model had earlier been rated (by a different group of women) as very good looking, the other as plain. The two models were dressed in one of three different costumes: a Burger King uniform, a plain off-white shirt, and a 'white dress shirt with designer paisley tie, a navy blazer thrown over the left shoulder, and a Rolex wristwatch showing on the left wrist' (ibid.: 376). The experimenters chose these costumes as indicators of low, medium and high socio-economic status respectively. The women in the study were shown pictures of each of the models, and then asked various questions about how willing they would be to perform various acts (such as having coffee and conversation, going on a date and getting married) with such a person. Scores were assigned to reflect their degree of willingness.

For the most part, the women Townsend and Levy interviewed expressed greater willingness to perform acts of all kinds when a model was in the high status costume than when the same model was dressed in the other costumes. What is more, when the plain model was dressed in the high status outfit, his scores were better for all acts than those of the handsome model dressed in the low status outfit. It might seem, then, that women care far more about status than looks. But there are problems with this interpretation. Buller points out that the handsome model in the medium status costume scored better than the plain model in the high status costume, which suggests that here, at least, looks are being given greater weight than status. More tellingly, Buller points out that all of the participants in the study were students from Syracuse University – a prestigious and expensive private university in the USA. If, as seems plausible, most of the Syracuse students interviewed consider their own socioeconomic status to be medium-to-high, then they may simply be picking mates whose perceived socioeconomic status is similar to their own. This would account for their general preference for models in medium or high status garb. Alternatively, the students might be trying to pick mates whom they regarded as having a similar educational background to their own. This preference, too, would account for their general lack of willingness to consider marrying anyone dressed in a Burger King uniform. In other words, the hypothesis that women's general preference is not for high status males, but for males who are in various respects similar to themselves, can also account for Townsend and Levy's data. This hypothesis is not mere conjecture. There are several sociological studies which suggest that marriage partners tend to be similar in a variety of respects, including socioeconomic status, ethnicity and religion.

What about male mate choice? As Buss notes, it is not clear precisely what preferences the evolutionary demands of our past environments predict, although he is confident that they predict some preference for youth. Perhaps men should be attracted to women of peak *fertility* – women most likely to conceive and give birth successfully. These are probably women in their early-to-mid-twenties. Or perhaps men should be attracted to women of peak *reproductive value* – women who, over the remainder of their lives, are likely to have the most offspring. These are probably women in their mid-teens. Do evolutionary considerations predict at least that all men should prefer women somewhere in the age range 15 to 25? Or should older men prefer women who are older than this? The answers are not clear.

One of Buss's studies looked at over 10,000 men and women, drawn from thirty-three countries on six continents (Buss 1989). On average, men in this group said they preferred to marry at 27.49 years of age. On average they said they preferred to marry women who were 2.66 years younger. So this does indeed mean that, on average, the men Buss surveyed preferred females who (at around 25 years old) were close to their peak fertility. What is more, while the men who participated in Buss's study preferred to marry women younger than they were, women in all countries tended to prefer men who were a few years older than themselves (3.42 years older, on average). So Buss has indeed provided evidence in favour of a widespread difference between male and female preferences. Even so, one cannot jump to the conclusion that men in general prefer women of peak fertility. The average

age of the men Buss surveyed was low – on average they were about 23.5 years old. Without information about the preferences of much older men one cannot rule out the hypothesis that men generally prefer women who, although a couple of years younger, are close to their own age. Other data sources suggest that human males do not generally prefer women of maximum fertility, only that they prefer women who are younger. Men in their fifties who remarry, for example, typically prefer women ten to twenty years younger (Buss 1999: 136). These women are well past maximum fertility. Buller also points out rather mischievously that not all men get divorced once their wives exceed their reproductive peak. This may be because these men would like to attach themselves to women in their early twenties, but feel they cannot for some reason. But it could also indicate that these older men have a preference for women of their own age.

Let me try to draw some conclusions from these studies. It seems likely that there are sex differences in mate preferences. Moreover, some of these differences may have an evolutionary explanation. But this does not mean that it is an easy matter to say what these preferences are, or precisely what form their explanation should take. Buss himself is aware of the complicated reality of mate preferences. Men do not all prefer women in their twenties, and it is not clear that all women put resources above looks. We need to guard against the temptation of taking a likely set of problems which humans faced in the past, and using these to argue that anyone who doubts that men prefer youth, or that women seek resources, is a head-in-the-sand anti-evolutionist. After all, if we place the stress on a different set of evolutionary problems, we can render quite different sets of mate preferences plausible from an evolutionary perspective. Evolutionary psychologists Kenrick and Keefe, for example, suggest that the need for a stable and effective alliance in order to raise offspring could have favoured preferences for a mate who is similar to oneself in various respects, including age:

> In comparison with other mammals, human males contribute a
> greater amount of care to their offspring. Although a preference
> for novel and young sexual partners would have contributed to a

> male's mating effort, any preference that led to bonding and cooperation with a mate would have contributed to parenting effort and increased fitness through the increased survival of off-spring. Extended interactions over long periods between mates would have been easier if the partners had similar expectations, values, activity levels, and habits. A preference for similarity in age, all else being equal, would have made the long-term cooperation of mates more feasible and thus adaptive.
>
> (Kenrick and Keefe 1992: 851)

It is easy to combine a picture of what 'everybody knows' about male and female mate preferences with a plausible adaptive scenario in our species' past, and to take this pair of views as unassailable evidence in favour of the deep-seated reality of some trait that is widely held to be part of universal human nature. But there is no guarantee that selection has acted in the ways that we initially judge most likely, and certainly there is every reason to be suspicious of the thought that selection formed us to accord with whatever the popular image of human nature happens to be. That is why we should be cautious before swallowing whole any claims about what the last twenty years of Darwinian psychology have taught us about what makes men and women tick, and about what policy makers need to know in order to do their jobs. And that is also why a proper evolutionary outlook on our species need not undermine the suspicion that claims about universal human nature sometimes say more about the dominant image we have of ourselves at a particular time than they do about the underlying features of our species that have persisted over time.

7. DARWIN AND THE LEFT

The prominent moral philosopher Peter Singer has argued that a proper Darwinian view of human nature makes some of the central themes of left-wing thinking untenable (Singer 1999). Singer's rough claim is that left-wingers have traditionally espoused the false views that human nature is infinitely malleable, and that the ills of society could be eradicated if only we educated people in such a way that they would lose their bad habits. Singer does not advocate abandoning left-wing goals (of reducing disparities in income, say, or of allocating resources according to individual need, rather than individual talent). But he argues that 'The Left' must rethink how its goals can realistically be achieved in the light of a Darwinian understanding of the reality and character of human nature. Hence he advocates a rethought 'Darwinian Left' which should not:

> Deny the existence of a human nature, nor insist that human nature is inherently good, nor that it is infinitely malleable;
>
> Expect to end all conflict and strife between human beings, whether by political revolution, social change, or better education;
>
> Assume that all inequalities are due to discrimination, prejudice, oppression or social conditioning. Some will be, but this cannot be assumed in every case . . .
>
> (Ibid.: 60–61)

My primary goal in the remainder of this chapter is to investigate the true political consequences of an evolutionary view of human nature. First, though, it is important to consider the relationship between the denial of the existence of human nature, and the assertion that human psychology is 'malleable'. Many researchers in sociology and anthropology put great store on the demonstration of human diversity. They argue that there is very little that humans have in common psychologically when we look across evolutionary time and space; moreover, they argue that when humans do have significant psychological features in common, this should normally be attributed to their common social or cultural backgrounds. It does not matter whether one equates this set of views with a denial of the existence of human nature, or with the view that human nature is real but changing and varied. What is important to note is that it does not follow from this set of views that the alteration of human psychology is an easy matter. Even if our psychological profile is socially determined – even if, for example, the tendency of young girls to want to play with dolls, and of young boys to want to play with guns, is caused by the exposure of girls and boys to different external stimuli (from parents, friends or other sources) – it does not follow that changing these tendencies (reversing them, or eliminating difference altogether) is a trivial affair. So even if human psychology is socially determined, and as a result of that is changeable, it does not follow that it is 'malleable' in the sense that a piece of metal in the hands of a good blacksmith is malleable. This also means that the general failure of 'social engineering' to change human behaviour does not show that scepticism about human nature (which Singer associates with the traditional Left) is misplaced.

Darwin himself was surprisingly open to claims about the malleability of human nature. As we saw, he thought natural selection explained the disparity in levels of male and female genius. But he was optimistic about the prospects for raising the levels of female genius to that of men. Darwin believed that faculties exercised during the life of an individual would tend to be passed on to offspring of the same sex. Men in the past needed to use their powers of thought far more than women, and so these powers were passed on to male offspring. Men needed to use these powers because females were so discriminating in their choices of mates that intelligence was a prerequisite in attracting a woman. It follows that the encouragement of women's powers of thought could bring them to parity with men:

> In order that woman should reach the same standard as man, she ought, when nearly adult, to be trained to energy and persever-ance, and to have her reason and imagination exercised to the highest point; and then she would probably transmit these quali-ties chiefly to her adult daughters.
>
> (*Descent*: 631)

Even so, Darwin points out that this would only improve the intel-lect of those women lucky enough to receive education and training (or to be descended from a well-educated woman). If the majority of women did not enjoy such education, and if this fact made no difference to their reproductive abilities compared with women who were educated, then we would see no gradual improvement over the generations of women considered as a whole. Finally, Darwin tempers his optimism regarding the lot of women with a final consideration that is sociological in character, for it relates to the demands that he perceived to be placed on men by day-to-day Victorian life:

> As remarked before of bodily strength, although men do not now fight for their wives, and this form of selection has passed away, yet during manhood, they generally undergo a severe struggle in order to maintain themselves and their families; and this will tend to keep up or even increase their mental powers, and, as a consequence, the present inequality between the sexes.
>
> (Ibid.: 631)

Or, in straight language, men remain more intelligent than women because daily life as head of a household requires that men make far greater use of their brains. It is a clear consequence of Darwin's views that inequalities between the intellectual endowments of men and women might be greatly reduced by making demanding education widely available to both sexes, and by encouraging women into positions of responsibility in working life. These remedies for inequality are, of course, wholly in tune with what Singer's traditional, non-Darwinian Left might recommend, even though Darwin takes the view that evolution is responsible for the inequalities in question.

A modern Darwinian might react to all this by saying that one who believes in the malleability of human nature gains no genuine support from Darwin's views, because Darwin's views on this matter are hopelessly wrong. Darwin believes, to repeat once again, that if a faculty is used during the life of an individual, that faculty is likely to be found in a developed form in the individual's offspring, too. Darwin's case for malleability appeals to the inheritance of acquired characteristics. He argues that men who exercised their creativity were at a reproductive advantage compared with men who did not, and that creativity was passed on to offspring by the mechanism of 'use-inheritance'. If women use their creativity too, then this faculty will be strengthened and then passed on to female offspring. The problem, our modern Darwinian might say, is that we no longer believe that acquired traits are inherited.

Now it is quite true that inheritance does not work in the way Darwin says it does (the question of whether there is any sense in which acquired traits are inherited will be raised in a moment). Does this mean that traits formed by natural selection are not malleable? It does not. Suppose, first of all, that human nature consists in a collection of genetically inherited adaptations. Would it follow from this fact that human nature is, as Cronin says, 'fixed'? On the view we are considering, natural selection builds adaptations when, over a number of generations, favourable differences in individual fitness, caused by genetic differences in the organisms that bear them, are added up and preserved. Of course genes do not produce organic traits all by themselves – what effect a gene has depends on the environment in which it is found (Lewontin 1985). So if, as we are assuming, these genetic differences produce fairly constant pheno-typic differences for the period during which the adaptation is constructed, then this must be because these genes are in fairly constant developmental environments. Hence alterations to develop-mental environments can change the effects genes have, which in turn is to say that they can result in the alteration of human adap-tations.

Now this argument is, admittedly, a little hasty. It is true that there is nothing in the logic of adaptation that tells us that adapta-tions are not malleable. Even so, developmental environments vary during the life of an individual and across generations. A fitness-enhancing trait which relied for its development on very finely specified environmental conditions would probably develop only rarely. Developmental processes are themselves subject to selection, and selection can 'buffer' development against environmental perturbations, thereby increasing the chances of fitness-enhancing traits emerging reliably in developing individ-uals in spite of local environmental changes. The consequence of this is that the development of genetically inherited adaptations is unlikely to be sensitive to environmental perturbations.

There are limits, though, to what natural selection can do.

Selection can make the development of an adaptation resilient to environmental perturbations that were characteristic of the environments in which the adaptation was shaped. But, as evolutionary psychologists Cosmides and Tooby rightly stress:

> Developmental processes have been selected to defend themselves against the ordinary kinds of environmental and genetic variability that were characteristic of the environment of evolutionary adaptedness, although not, of course, against evolutionarily novel or unusual manipulations.
>
> (Tooby and Cosmides 1992: 81)

Selection in our species' past did not have the foresight to make the development of our adaptations resistant to environmental alterations that have emerged only in modern society. Neither did it need to make development resistant to environmental alterations that existed only rarely in the past. Even if human nature is a suite of genetically-inherited psychological adaptations, we should still expect nature to be subject to alteration when it meets new or rare environments, both of which have the potential to derail the development of those adaptations.

Senator J. William Fulbright is alleged to have said, at the Man and Beast conference in Washington, DC in 1969:

> If we assume that men generally are inherently aggressive in their tendency, . . . if this is inherent and man cannot be educated away from it, it certainly makes a great deal of difference in one's attitudes towards current problems . . . If we are inherently committed by nature to this aggressive tendency to fight, well then, I certainly would not be bothering about all this business of arms limitations or talks with the Russians.
>
> (Quoted in Segerstråle 2000: 92)

We have seen that there is no good argument from the image of human nature as a set of genetically-inherited adaptations to the image of human nature as fixed. Does this mean we should dismiss Fulbright's worry? The fact that adaptations are, in principle, subject to alteration via manipulation of the developmental environment does not entail that the specific form of developmental manipulation they are most susceptible to is alteration of the educational environment. Development is not well understood, and the ways in which early developmental interventions can affect the appearance of later traits can be highly counterintuitive. So if aggression is a genetically-inherited adaptation, then nothing we have said so far establishes that it will go away by educating people differently. If we put a bet not merely on human nature being subject to alteration, but on the malleability of human nature by education, then we are sticking our necks out way beyond what the evidence demands.

This, I think, is a fair point. Stepping back from psychological adaptations, we would be unlikely to argue that changing educational regimes will make a difference to our anatomical or physiological adaptations. Perhaps we can teach people to walk in different ways, so that their gait is improved, but teaching people in different ways will not change the colour of their eyes (although even now tinted contacts lenses allow us to change the apparent colour of our eyes with ease). Some of our dispositions of thought might also be impervious to changes in how we are taught. Only on a case-by-case basis could we determine whether educational interventions also make a difference to specific cognitive adaptations. But Singer's traditional Left should gain hope from the case made in chapter seven regarding the adaptive advantages of the ability to learn itself. If natural selection has built our species to be receptive to learning, then very many psychological traits will indeed reflect what recent generations and the changing environment teach us, in ways that can be unlearned as the social, physical and biological environments change.

Up to now I have been assuming that any adaptations that natural selection builds are inherited in virtue of the transmission of genes. But natural selection works whenever organisms in the parental generation differ in ways that affect their fitness, so long as offspring resemble their parents in terms of those differences. We usually think genetic differences between organisms in the parental generation explain why offspring inherit the distinguishing characteristics of their parents. Yet this need not be the case (Mameli 2004). To take just three examples, people differ in how wealthy they are, and they pass their wealth on to their offspring. People who are well-educated often seek a good education for their children, and the result is that their children become well-educated too. People also differ in their moral codes, and they influence the moral views of their offspring by direct teaching, and more indirectly by influencing the environments in which their children learn. If an individual's level of wealth, or of education, or their moral outlook, have systematic effects on their prospects for survival and reproduction (for example, by improving access to healthcare, or facilitating the attraction of mates), then natural selection can act on fitness differences that are not explained by genetic differences. What is more, wealth, education and moral fibre are the kinds of properties that can be acquired during the life of an individual, and then passed on to the individual's offspring. So it is not quite true to say that modern biology rules out the possibility of the inheritance of acquired traits.

We should not dismiss the thought that forms of non-genetic inheritance, including social and cultural inheritance, have been important in the evolution of our species. If these forms of inheritance have been important, then our species' adaptations are likely to be subject to alteration by manipulating the social structures which enable their reliable development across generations. Perhaps the most interesting form of non-genetic inheritance in this context is based on *niche-construction*. As Darwin recognised, organisms do not just adapt to fit an independently structured environment, they transform and maintain their environments in an active way. This is one of the themes that Darwin stresses in his last published work, a text on earthworms. Even worms have the capacity to modify their surroundings, so much so that large boulders can be buried or lifted by the repeated laying down of worm castes. Niche-construction can play a role in inheritance, because the parental generation can act in ways that ensure the maintenance of the environment in which their offspring will develop (Sterelny 2001a). Adult beavers maintain a watery environment by the construction of dams. This environment facilitates the acquisition of dam-building skills by young beavers. Niche-construction processes partly explain resemblances between parents and offspring, and they do so in virtue of the collective action of the parental generation, rather than in virtue of something (genes, wealth or knowledge) passed from a single parent to its offspring. Humans are masters of niche-construction. We collectively build environmental features such as libraries, schools and hospitals, which in turn facilitate the development of future generations, ensuring that they possess the skills and abilities to renew these same libraries, schools and hospitals.

I began this chapter by noting that Darwin's theory is some-times associated with the political views of the hard Right. Karl Marx was also attracted to Darwin's theory, writing in 1861 to Ferdinand Lassalle that 'Darwin's work is most important and suits my purpose in that it provides a basis in natural science for the historical class struggle' (Marx and Engels 1936: 126). Near the beginning of his short book, Singer gives a quotation from Marx's *Theses on Feuerbach* as an example of the kind of view he thinks the Left needs to abandon: ' . . . the human essence is no abstraction inherent in each single individual. In its reality it is the ensemble of the social relations' (Singer 1999: 5).

Darwin does not prove Marx right. On the other hand, it is not clear to me that a good Darwinian must drop Marx's view of human nature. We should not think that human nature is an abstraction inherent in each single individual, at least not if that means human nature is a kind of genetically-specified core which we all share. Commonly-held genes develop to yield commonly-held psychologies in the context of common developmental environments, structured and maintained by social interaction. Social and cultural inheritance systems contribute to the reliable development of human psychology not only when individual offspring learn from their parents, but also through the collective action of whole generations to maintain robustly-structured devel-opmental environments for the generations to come. A good Darwinian still has plenty of reasons to think that 'the ensemble of the social relations' plays an important role in the construction and re-construction of human nature.

SUMMARY

Darwin's applications of his own theory to political matters may wrong-foot the modern reader. For example, although he worries that our tendency to care for the sick is likely to lead to degeneration in the species, he does not advocate that we cease this care, for he thinks to do so would have even worse consequences than degeneration itself. Although he believes that natural selection has made men greater geniuses than women, he thinks that demanding educational regimes can nonetheless lift the genius of individual women to that of men. Modern Darwinians rarely attempt to link evolutionary ideas to political issues, but when they do, it is usually in quite different ways to those of Darwin. Sometimes, they argue that good political interventions require an understanding of human nature, and that evolutionary psychology is the science to provide such an understanding. In principle this is perfectly legitimate, but we must be sure to subject the specific claims that issue from evolutionary psychology about the nature of human nature to scrutiny. More rarely, modern Darwinians argue that many left-wing thinkers are committed to an untenable scepticism regarding the fixity of human nature. But there are two reasons for thinking these left wingers need not alter their stances in the face of evolutionary research. First, even if we accept that human nature consists in a suite of genetically-inherited adaptations, such adaptations are still subject to alteration. Second, it is a mistake to think that genes are the only significant agents of human inheritance. In a variety of ways, social and cultural resources play roles in inheritance, too, and by extension they are implicated in human nature.

FURTHER READING

Once again, *Descent* is the best place to look for Darwin's own sustained discussions of matters political, especially chapter five (on the effect of selection on civilised nations), chapter seven (on race) and chapters nineteen and twenty (on differences between men and women). The introduction to *Descent* by Desmond and Moore includes a detailed discussion of Darwin's views on race.

For a history of eugenics, Daniel Kevles's book is the best place to start:

Kevles, D. (1995) *In the Name of Eugenics*, Cambridge, MA: Harvard University Press.

The relationship between Darwin's views and eugenics can be found in:

Paul, D. (2003) 'Darwin, Social Darwinism, and Eugenics', in J. Hodge and G. Radick (eds) *The Cambridge Companion to Darwin*, Cambridge: Cambridge University Press.

Evelleen Richards has written an impressive historical study of Darwin's views on women:

Richards, E. (1983) 'Darwin and the Descent of Women', in D. Oldroyd and I. Langham (eds) *The Wider Domain of Evolutionary Thought*, Dordrecht: Reidel.

Political allegations of one kind or another were slung around with frequency during the sociobiology debate of the 1970s. There is a lively history of this debate:

Segerstråle, U. (2000) *Defenders of the Truth*, Oxford: Oxford University Press.

Peter Singer makes the case for the Left's need to rethink some of its assumptions in a short and snappy book:

Singer, P. (1999) *A Darwinian Left: Politics, Evolution and Cooperation*, London: Weidenfeld and Nicolson.

David Buller considers sex differences and various other claims about human nature relevant to policy in detail:

Buller, D. (2005) *Adapting Minds: Evolutionary Psychology and the Persistent Quest for Human Nature*, Cambridge, MA: MIT Press.

The case for integrating niche-construction into human evolution is made in a variety of places, including:

Sterelny, K. (2003a) *Thought in a Hostile World: The Evolution of Human Cognition*, Oxford: Blackwell.

Nine

Philosophy

I. MAN'S PLACE IN NATURE

At the most abstract level, Darwin's gift to philosophy is the gift of genealogy. By showing decisively that the human species is a part of the Tree of Life, Darwin encourages us to study ourselves in the same way we would study any other species. We should see the human capacities which have fascinated philosophers – the capacity to praise, to blame, to be moved, to cooperate, to know, to plan and to act – as the products of historical processes, capacities whose functions have been modified over time, and which still bear the marks of earlier roles. These capacities have shaped our physical, biological and social environments, and they have been shaped by those environments. In these general respects Darwin's views are close to those of the nineteenth century's other great genealogist, Friedrich Nietzsche.

Human capacities are the products of historical processes, but which processes? Darwin regarded natural selection as the most important influence on the makeup of the organic world in general, but he did not claim that human capacities should be explained exclusively in terms of natural selection, and he did not understand natural selection in quite the way we do today. His discussions of the moral sense and the emotions place natural-historical explanation in the foreground, but he supplements natural selection with distinctive appeals to use-inheritance, reasoning, experience and learning in accounting for the origin and continued modification of these faculties. In these respects, Darwin was no Darwinian.

In this concluding chapter I want to extend these general reflections on Darwin's philosophical impact. Darwin teaches us that
life is a great tree, and our species is a twig on that tree. We can
take comfort from this image, stressing as it does our kinship not
just with apes, but ultimately with all of nature. Yet Darwin's view
also exposes, in the eyes of many, a kind of hubris that our species
has been prone to. We are not distinct from nature, we do not
ride above it, and neither are we evolution's greatest work. We,
along with worms and insects, are elements of life's 'entangled
bank', which Darwin ruminates on at the very end of the Origin:

> It is interesting to contemplate an entangled bank, clothed with many
> plants of many kinds, with birds singing on the bushes, with var-
> ious insects flitting about, and with worms crawling through the
> damp earth, and to reflect that these elaborately constructed
> forms, so different from each other, and dependent on each other
> in so complex a manner, have all been produced by laws acting
> around us.
>
> (*Origin*: 459)

The continuation of this passage draws the Origin to a close, and
casts some doubt on the thought that in Darwin's eyes Homo *sapiens*
is nothing special:

> Thus, from the war of nature, from famine and death, the most
> exalted object which we are capable of conceiving, namely, the
> production of the higher animals, directly follows. There is
> grandeur in this view of life, with its several powers, having been
> originally breathed into a few forms or into one; and that, whilst
> this planet has gone cycling on according to the fixed law of
> gravity, from so simple a beginning endless forms most beautiful
> and most wonderful have been, and are being, evolved.
>
> (Ibid.: 459–60)

The 'higher animals' are 'the most exalted object we are capable
of conceiving': does Darwin think our species, although a part of
nature, is the highest animal of all?

2. HUBRIS

Darwin is sometimes portrayed as one of a series of revolutionary thinkers who have exposed the modesty of man's position. Copernicus demonstrated that the Earth is not at the centre of the Universe, but merely one of many planets revolving around the Sun. Darwin shows that Man is not a species apart from nature or above it, but, like all species, one among many of the branches of the tree of life. Darwin shows us that our self-image needs cutting down to size. Our species is unique, but uniqueness is ubiquitous in nature.

One could try to save man's privileged place in nature not by denying our kinship with other species, but instead by showing that, in some sense, we are the culmination of evolution's work, the grand crescendo that all these billions of years have been working towards. But this, too, is a view that many think Darwin has successfully destroyed. Stephen Gould, in particular, has sought to undermine two views that form the picture of man as the icing on the evolutionary cake. First, he has argued that evolution is a highly contingent affair (Gould 1991). The species that we see on Earth, including our own, are by no means the necessary results of evolutionary processes. The outcomes of evolution, Gould thinks, are highly sensitive to small perturbations and slight disturbances, with the result that, were we to 'replay life's tape', we would see quite different results each time. Second, Gould argues (along with many other modern biologists) that evolution is not progressive (Gould 1996). When Microsoft tells us that Office has 'evolved', they presumably intend this to carry a connotation not merely that its products have changed, but that they have changed for the better. Gould thinks that in biology, at least, it is grossly misleading to give evolution this progressive reading. Gould believes that humans are produced by evolution, but our existence is a fragile outcome of a process with no tendency towards improvement.

The notions of necessity and progress, while united in a view of man as the inevitable high-point of evolution, are different. If there are many different ways for life to get better, and if very small perturbations can lead to evolution taking one upward route rather than another, then the history of life might be both progressive and contingent. Conversely, the history of life might be highly constrained, so that the same evolutionary end-point is likely to be attained no matter what the starting conditions. Even so, that end-point might be no better than the various possible starting positions. Here we have no progress, but no contingency either. The two ideas need to be investigated separately.

3. CONTINGENCY

What does Gould have in mind when he says that evolution is highly contingent? He is not making a claim about how likely it is for life to have emerged given the general features of our planet, nor is he making a claim about how likely it is for a planet to have features conducive to life. He is not merely saying that large environmental catastrophes (supernovae, meteorite strikes, magnetic pole reversals) can lead to mass extinctions and new evolutionary consequences for the survivors. Gould's idea is, roughly, that once evolutionary processes get going, very small changes become magnified quickly over time. Had the history of life been even slightly different, the makeup of the world's species today would be completely different.

The claim that small, chance events can affect which species come to exist is a conclusion that can be made very easily by those who view species as *individuals* (see chapter three, section four). We can see this by considering an analogous contingency claim about individuals of a different kind – namely, individual people. It is reasonably plausible to think that had my parents never met, or had they met a little earlier, or a little later, then I would not have existed, even though someone (or maybe a few people) rather like me might have existed. Now consider the view that new species are formed by geographical isolation. Take a species that is originally formed when a small group of birds becomes detached from the main flock, and is blown off course onto an island. Some versions of the species-as-individuals view will say that had the island been settled by a different group of birds separated from the same flock, then the resulting species would not have been the very same species as the one we in fact see. This is a cheap way to secure the contingency claim, for although it suggests that if the tape of life were replayed, we would be unlikely to see the emergence of Homo sapiens, it leaves open the thought that replaying the tape would always yield species very similar to our own.

Gould's claim is an interesting one if it says not merely that the human species is contingent, but that humanoids – organisms similar to us in many respects, both internal and external – are contingent (Sterelny 2001b). As we have seen, the traits a species acquires over time depend on (among other things) the variation available in that species, and on the environmental demands made on the species. Theoretically, both of these factors can be sensitive to small perturbations, hence the adaptations a species acquires can be sensitive to small perturbations, too. Small variations in flight pattern might make a difference to which group of birds is detached from the main flock. The question of which small group ends up isolated on a remote island can make a difference to the range of variation that is available for selection to act on as the new species is formed, for the genetic makeup of the breakaway group may be thoroughly atypical in comparison to the parent group. Or consider the action of sexual selection. There is considerable evolutionary advantage to be gained from having offspring whose physical appearance conforms to the mating preferences of other members of the species – your fitness is augmented if you have sexy sons and daughters, because this makes your sons and daughters more likely to reproduce. But if those mating preferences (blue feathers or yellow? a long tail or a tall crest? pale skin or dark?) are initially determined by chance, the adaptive evolution of the species as a whole can be dependent on minor chance occurrences.

These are primarily theoretical considerations in favour of Gould's contingency claim. One might counter them with more theoretical arguments. Even if some sexually-selected traits are subject to contingency, perhaps many other distinctive humanoid traits – most obviously the abilities to plan and to communicate that underlie human intelligence – are of a type that will provide advantages to survival and reproduction across a very broad range of different environments. These are traits that are likely to evolve no matter what.

Theoretical considerations can be brought to bear on both sides of the argument. Ideally, we should resolve these theoretical disputes with an empirical test of the contingency claim. The problem is that we cannot literally re-run the tape of life to see whether humans usually evolve. Even so, the biologist Simon Conway Morris (ironically, one of the men whose work Gould uses to back his contingency claim) has argued that what we know about evolution counts against Gould (Conway Morris 2003). Conway Morris argues that *convergent evolution* is a regular occurrence on Earth. Evolution is said to be convergent when similarly structured traits arise independently in the face of similar environmental demands. The wings of birds and bats are an example of evolutionary convergence, because birds and bats have not inherited their wings from a common winged ancestor, but have acquired these similar traits independently, because of the demands placed on both by flight. What is the link between convergence and contingency? Birds and bats are different – they have different histories, different anatomies, their environments are not identical. Even so, roughly similar environments led them to evolve roughly similar wings. This is indeed empirical evidence against the claim that small differences in evolutionary histories make big differences to evolutionary outcomes. If small differences mattered as much as Gould says they do, then convergence like this should be very rare.

Conway Morris's catalogue of convergences is suggestive, but the philosopher Kim Sterelny argues that it does not suffice to undermine Gould (Sterelny 2001b). One thing to note is a limitation in the scope of Conway Morris's argument. His evidence suggests that perhaps we could re-run parts of the Earth's tape of life, and humanoids would still appear. That does not mean that we should expect to find humanoids on any planet where life evolves. Remember that bats and birds have common ancestors, and they share a great many properties in virtue of that. Such properties include tendencies to produce some kinds of variation rather than others. And these tendencies, held in virtue of common ancestry, may be part of the explanation for convergence. This makes it hard to assess what would happen if life emerged on an alien planet, for the organisms on such a planet would not share common ancestors with us, and the evolutionary pathways they follow might be quite different as a result. Second, the existence of a few cases of convergence is compatible with Gould's contingency claim. For even if evolution were sensitive to small perturbations, we should still expect to see instances of convergence every now and then. What Gould cannot endorse is the claim that convergence happens more often than not. And here we run into another problem. We can only show that convergence is the norm if we have some way of counting how often convergence does not occur. But it is far from clear how we could do that in a sensible way. Is our own species' failure to sprout wings an instance of failed convergence on the grounds that it would be good for us to take to the air? Presumably not, but unless we have some principled way of calculating a ratio of convergence to non-convergence it is unclear how to decide one way or the other about contingency.

4. PROGRESS

Darwin says little about contingency, but he says a good deal about progress. What he says points in different directions, in part reflecting the difficulty of the topic (Shanahan 2004). To say that evolution has shown progress is to say that there has been a change over evolutionary time, and that it is a change for the better. Things could get better over time purely by accident, so to say that evolution is progressive also carries an implication that better forms are reliably favoured by evolutionary processes.

In his notebooks Darwin expresses a good deal of suspicion regarding the criteria we might use to say that improvement has been made: 'It is absurd to talk of one animal being higher than another.—*We* consider those, where the cerebral structure/intellectual faculties most developed, as highest.—A bee doubtless would where the instincts were.—' (B Notebook, quoted in Barrett et al. 1987: 189).

Now there are creatures with large brains; before there were not. But it is only the fact that we have those large brains that makes us count this situation as higher than what went before: 'Why is thought, being a secretion of the brain, more wonderful than gravity a property of matter? It is our arrogance, it our admiration for ourselves.—' (C Notebook, quoted from ibid. 1987: 291).

On this view, evolutionary history is one of change that is, in itself, flat; it is only our self-regard that causes us to project an upward trend onto this history. Gould has seized on these remarks and others to argue that Darwin rejected the linkage of evolution and progress, and Gould is in no doubt about the wisdom of this rejection: 'Progress is a noxious, culturally embedded, untestable, nonoperational, intractable idea that must be replaced if we wish to understand the patterns of history' (Gould 1988: 319).

Noxious or not, Darwin was not wholly opposed to the idea of progress, and he conquered, at least partially, his early doubts about criteria for higher and lower. He held the view that the evolutionary record does manifest progress, in two senses that are not dependent on using ourselves as a standard of value (Shanahan 2004). First, Darwin claims that later organisms are better suited in the struggle for life than the earlier organisms which they supplanted:

> The inhabitants of each successive period in the world's history have beaten their predecessors in the race for life, and are, in so far, higher in the scale of nature; and this may account for that vague yet ill-defined sentiment, felt by many palaeontologists, that organisation on the whole has progressed.
>
> (*Origin*: 343)

Second, Darwin thinks that later organisms show greater speciali-sation in their parts than earlier organisms. He uses increasing division of labour within an organism to measure progress, a measure which he inherits from Adam Smith via the biologists Henri Milne Edwards and Karl Ernst von Baer. Just as economic progress consists in the increasing specialisation of workers' roles, so biological progress consists in the move from single-celled organisms with little internal differentiation, and whose parts play many diverse roles, to creatures with specialised tissues and a multitude of different organs (a heart for pumping blood, a brain for thought, limbs for locomotion). Darwin believes that there has been progress, understood as an increase in complexity, and he takes the specialisation of parts to be a measure of complexity.

Finally, Darwin believes that natural selection links the concepts of progress as competitive advantage and progress as specialisation. An organism whose parts are specialised to particular roles will tend, says Darwin, to be more efficient overall than an organism whose parts execute many different functions. Hence natural selection favours an increase in complexity over time. For Darwin, selection is indeed a progressive force in evolution, as expressed in this passage (which appears in the second edition of the *Origin*, published only a month after the first), where a definition of 'high' is put forward:

> The best definition probably is, that the higher forms have their organs more distinctly specialised for different functions; as such division of physiological labour seems to be an advantage to each being, natural selection will constantly tend in so far to make the later and more modified forms higher than their earlier progenitors, or than the slightly modified descendants of such progenitors.
>
> (Darwin 1959: 547)

If Darwin's view of progress is the right one, it still falls far short of endorsing a view of our own species as sitting on top of the evolutionary tree. Darwin regards progress as a general pattern across time: natural selection has a tendency to favour an increase in complexity, so today's species are, in general, higher than their ancestors. At best this may allow us to rank ourselves more highly than the single-celled organisms from which we are descended, but it says nothing about whether, for example, today's humans are higher than today's termites. This may explain why in the

Origin's third edition, Darwin continues to worry about comparing very different species in terms of their level of advancement, even after he has arrived at definitions of 'higher' and 'lower' that he is happy with:

> To attempt to compare members of distinct types in the scale of highness seems hopeless; who will decide whether a cuttlefish be higher than a bee, that insect which the great Von Baer believed to be 'in fact more highly organised than a fish, although upon another type'?
>
> (Ibid. 1959: 550)

Progress in Darwin's sense is produced by the action of external conditions on blind variation. It is likely to occur only in so far as increasing complexity confers an advantage in the struggle for existence, and only in so far as variation happens to arise that is more complex. Darwin thinks it generally likely that such conditions will be met, but there is no assurance. Hence he notes that on some occasions an increase in complexity will be detrimental:

> . . . we can see . . . that it is quite possible for natural selection gradually to fit an organic being to a situation in which several organs would be superfluous and useless: in such cases there might be retrogression in the scale of organisation.
>
> (Ibid. 1959: 222)

There is no intrinsic drive to improvement that resides within organic nature itself. Improvement is indeed the tendency we should generally expect in evolution, but this comes from without, not from within, and it is subject to stall when outward conditions are not propitious. I take it that this is what Darwin is stressing in the Origin's third edition when he distances himself from Lamarck, 'who believed in an innate and inevitable tendency towards perfection in all organic beings' (ibid.: 223).

How plausible is Darwin's equation of progress with competitive advantage in the struggle for existence? Selection occurs when fitter forms replace those which are less fit. It is, indeed, part of the theory of natural selection that those organisms which are at a competitive advantage will tend to do better. This, in part, is what Richard Dawkins has in mind when he writes that:

> . . . adaptive evolution is not just incidentally progressive, it is deeply, dyed-in-the-wool, indispensably progressive. It is fundamentally necessary that it should be progressive if Darwinian natural selection is to perform the explanatory role in our world view that we require of it, and that it alone can perform.
>
> (Dawkins 1997: 1017)

That role is the explanation of complex adaptation by the gradual accumulation of progressively fitter variants. Even so, this is a minimal sense of progress, because it does not deliver as much as we might think.

If fitter forms replace less fit forms, we might think that this will lead to a necessary increase in the average fitness of a population. But this is not so, as the problem of 'subversion from within' illustrates. Take the case we saw in chapter six, of a group of altruistic organisms that take turns to patrol their territory, issuing warning signs when a predator is detected. Now suppose a loafer – one who does not take turns to patrol – arises in the population by mutation. The loafer is fitter than the patrollers, because the loafer gains the benefits of the patrollers' warning signs without paying the cost of patrolling and its consequent dangers. So loafers will out-compete patrollers until (let us suppose) we move from a group with 100% patrollers to a group with 100% loafers. When the group was full of patrollers, the individuals in it were able to evade predators. This will no longer be the case when it is composed wholly of idle loafers. So this is a change brought about by selection of the fitter organism, but it is hard to view it as progress, because the average fitness of a group of loafers will be lower than the average fitness of a group of patrollers (Sober 1993).

Dawkins is not interested in defending an increase in fitness as the criterion of progress. As he recognises, many evolutionary situations constitute 'arms races', in which, for example, the running speeds of both predators and prey increase in tandem, with the result that there is no general improvement in the survival and reproduction of either species over time. On Dawkins' view, what increases is *adaptedness*: fast runners may reproduce at just the same rate as earlier slow runners, but they are better adapted. How are we to say when one organism is better adapted than another? One way to do it is Darwin's way: organism *a* is better adapted than organism *b* if *a* would tend to out-compete *b*. We are right to say that zebras have improved over time as their running speed increases because, although the average fitness of a herd of fast runners may be the same as the average fitness of an earlier herd of slow runners, if they were put together, the fast runners would be at a competitive advantage. But this leads to a second problem. The notion of competitive advantage is not *transitive*. That is, if *a* out-competes *b*, and *b* out-competes *c*, it does not follow that *a* out-competes *c*. As Kim Sterelny puts it, some evolutionary scenarios are like scissor– paper–stone games: paper beats stone, stone beats scissors, but paper loses to scissors (Sterelny 2001b). This means that in some species, although today's organisms out-compete yesterday's, and yesterday's out-compete the day before's, today's organisms would lose to the day before yesterday's. A pattern of competition just like this has been observed in successive strains of yeast (Paquin and Adams 1983). Natural selection makes it likely that the members of a species are at a competitive advantage compared only with their own immediate predecessors, not compared with all of their earlier ancestors. So once again, it is somewhat misleading to say that natural selection has a tendency to bring about progress.

Finally, the equation of the higher form with that which has a competitive advantage only allows us to make very restricted comparisons, because we can only compare organisms that are engaged in the same type of competition. Perhaps it makes sense to ask whether Roger Federer's game represents an improvement over John McEnroe's. This is to ask who would win an imaginary tennis match between McEnroe in his prime, and Federer in his. Of course finding out the answer to this question is impossible, but at least the question is intelligible, unlike the question of whether David Beckham's play represents an improvement over McEnroe's. What would it mean to compare these two? Are we supposed to imagine them playing tennis with a football? Similarly, while we might be able to compare fast running zebras with slow running zebras, it makes no sense to ask whether zebras would out-compete oak trees, or whether cuttlefish would out-compete algae: these organisms are not in the same game.

Darwin's criterion of progress as competitive advantage, and Dawkins' interpretation of progress as an increase in adaptedness, allow only local claims of improvement which compare members of a species with their own immediate ancestors. Neither allows us to regard the general trend of evolution as one from lower to higher. Even so it is hard to resist the thought that, in some sense, things have moved upwards in the last three and a half billion years. Back then, prokaryotes were the only living things to be found on Earth. These are single-celled organisms, which do not possess a nucleus. Prokaryotes are still around now, and in massive numbers, but today there are also multi-celled plants, social insects, birds with abilities to make tools, dolphins with abilities to communicate. Is there any sense in which Darwin is right to think that there is a trend towards increasing complexity?

Gould thinks there is such a trend, but only because when life starts it is as simple as it can be (Gould 1996). If he is right, then life can only increase in complexity, regardless of whether complexity is actively favoured by evolution. Sterelny points to a meatier notion of progress, which builds on the work of biologists John Maynard Smith and Eors Szathmary (Sterelny 2001b). They have proposed a view of evolutionary history as characterised by a series of 'major transitions' (Maynard Smith and Szathmary 1995). Their view of how biological complexity has increased over time is based, like Darwin's, on a conviction that the division of labour brings increasing efficiency. Specifically, they claim that the history of life has been punctuated by the development of new inheritance systems – what they describe as new modes of 'information storage' – which are specialised systems for bringing about reliable reproduction. They argue that DNA, packaged in genes on nuclear chromosomes, constitutes only one such system for information storage, one that permits faithful reproduction and thereby facilitates the construction over time of complex adaptations.

Here, in edited and abridged form, are a few of the transitions they recognise. One is from prokaryotes to eukaryotes. Prokaryotes, as we have seen, are single-celled organisms, and the cells in question are rather rudimentary. Eukaryotes are also single-celled organisms, but they, unlike prokaryotes, have nuclei, and also structures outside the nucleus called mitochondria. It is widely believed that the mitochondria of eukaryotic cells were initially free-living prokaryotes. A second transition is from single-celled eukaryotes to multi-cellular plants and animals, which are composed of many eukaryotic cells. Maynard-Smith and Szathmary also argue that a transition occurred with the advent of social organisms; specifically, they argue that groups with a strong division of labour between different types of individual, such as colonies of ants or termites, can be understood as organisms in their own right.

In each of these transitions we see a coming-together of entities that once had the potential to live and reproduce independently, but which have become united in a greater collective that has the ability to reproduce in its own right. Prokaryotes lose their reproductive autonomy when they become parts (the mitochondria) of single-celled eukaryotes. Eukaryotic cells lose their autonomy when they become parts of specialised organs in multi-celled animals or plants. Animals lose their autonomy when they become specialised individuals (sterile worker bees, for example) in larger colonies.

Might there be a general trend towards the formation of increasingly elaborate collectives, new types of organism whose parts were once organisms in their own right? If so, what might explain that trend? If Darwin is right about the advantages of division of labour, then a collective whose members perform specialised roles is likely to produce more offspring than a collective of generalists. But remember the problem of subversion from within. Selection within a collective, where members of the collective compete with each other, will often disrupt the organisation of the collective as a whole. Consider the collection of cells that is an individual animal. Compared with other cells in the animal, cancer cells are very fit. But their swift rate of spread compared with other cells in the body can result in the death of the animal that houses them. Different levels of selection interact (Buss 1987): selection acting on individual animals (for example) favours the repression of the ability of individual cells within those animals to go it alone by becoming cancerous. 'Major transitions', on this view, happen when selection at the higher level of reproducing collectives causes both a reduction in the ability of lower-level elements to reproduce independently, and an increase in the adaptation of these elements to specialised roles within the collective, thereby securing (for the higher-level units) the advantages of division of labour against the threat of subversion from within.

To show that a series of major transitions is a reliable evolutionary trend, rather than a mere fact, we would have to show not only that there have been several transitions of the kind sketched above, but that higher levels of selection have a general tendency to dominate over lower levels. This will not be an easy job. Even if it turns out that there is such a trend, it would not mean that our own social species is evolution's high point; only that evolution has a tendency to produce ever more comprehensive collectives. Putting this issue aside, work on transitions yields an interesting perspective on the question of what attitude a Darwinian should have to human nature. The organisation of humans into social groups seems to qualify as a major transition; one might argue that the organisation of social groups into integrated polities is another. Maynard Smith and Szathmary argue that human language is a new mode of information storage, one which opens up new kinds of evolutionary possibility. Regardless of how we feel about understanding this transition in terms of information storage, the definition of a transition in terms of the reduction of reproductive independence of parts seems to hold good when it comes to human socialisation and knowledge exchange. Human couples have a very limited ability to reproduce independently. The chances of two parents raising a baby to maturity with literally no recourse to the assistance of others – whether that is in terms of medical care or the provision of food, shelter, education and so forth – are low. Humans are reliant for survival on an array of social resources; sometimes directly, when others provide goods or perform services for us, other times indirectly, when we help ourselves using knowledge gleaned from others. Complex human societies, in which labour is divided in a variety of ways, are renewed each generation through the interaction of diverse cultural and biological resources; meanwhile, the ability of human individuals to reproduce independently of social assistance is severely limited. Far from encouraging a purely biological perspective on the human condition, Darwin encourages us to see human society in organic terms, and the human organism in social terms.

5. DARWINIAN NATURALISM

It is time to draw this book to a close. What morals can we draw for how to do philosophy from the success and fertility of Darwin's work? Many philosophers have argued that they must become more engaged with natural science if their subject is to make advances. At the beginning of chapter one we noted Darwin's own disappointment with his grandfather's evolutionary theory, 'the proportion of speculation being so large to the facts given' (*Autobiography*: 24), and with the 'deductive manner' of Herbert Spencer's philosophising (ibid.: 64) – that is, a style of philosophy which pays insufficient attention to facts, and which relies too much on the abstract tools of logic. Darwin's writings – which ask questions of the most ambitious sort regarding morals, the mind, knowledge and politics – show that he is not opposed to philosophical engagement, only to a particular speculative form of philosophical method. Darwin's theoretical ambition, directed and corrected by close attention to the results of numerous scientific disciplines, makes his approach to asking questions of the world a methodological model for many modern Darwinian philosophers. These philosophers often attribute the widespread sense that philosophy failed to make progress in the mid part of the twentieth century to that period's obsession with armchair theorising regarding the meanings of various concepts – truth, belief, emotion, right and wrong – at the expense of active empirical enquiry. These Darwinian enthusiasts advocate a view that we can call 'Darwinian Naturalism': 'Naturalism' because they say that philosophy needs to engage with, or even become a part of, natural science, and 'Darwinian' because it is the areas of science in which Darwin worked which they say have most relevance to philosophical concerns.

Naturalism is a position which advocates some form of deference among philosophers to our best scientific theories. Naturalism therefore comes in various strengths. In its weakest forms, it says only that whatever philosophical views we hold, they should not contradict our best science. This is a weak naturalism, because it tells philosophers not to tread on the toes of scientists, without advocating any constructive role for science in doing philosophy. The philosopher Alexander Rosenberg gives a stronger account of naturalism:

> Among philosophers, naturalism is the view that contemporary scientific theory is the source of solutions to philosophical problems. Naturalists look to the theory of natural selection as a primary resource in coming to solve philosophical problems raised by human affairs in particular.
>
> (Rosenberg 2003: 310)

We have seen plenty of examples in this book of constructive appeals to Darwin: the invocation of natural selection as a way of making innate knowledge respectable for the empiricist is an example. At the limit, one can imagine a kind of ultra-naturalism, which says that science is the sole repository of answers to problems that we have hitherto regarded as philosophical. The ultra-naturalist says that abstract philosophical reflection is a waste of time: questions regarding the nature of knowledge, or of a life worth living, should be answered directly by hard-headed empirical investigation, not by scratching one's beard in the bath. Darwin himself strikes me as a strong naturalist, but not an ultra-naturalist: he believes that natural historical reflections can raise the level of philosophical work done on morality, for example, but I see little evidence that he thinks traditional philosophy might be abandoned altogether. The reason why we should not, and could not, give up entirely on philosophical reflection of a more abstract kind is that the consequences of scientific results for philosophical problems are rarely transparent. We have seen this time and time again in the course of this book. Consider ethics: does evolutionary theory tell us that there are no ethical facts, or does it instead tell us that there are ethical facts, and that they are wholly 'natural'? The problem here is that to answer this question we need to be clear on what sorts of things we think facts are, what sorts of things values are, and so forth. And it is in getting clear on matters like these that much philosophy gets done – philosophy that might on occasions be informed by the natural sciences, but which cannot read its answers off from the sciences' results.

So far we have been considering naturalism in general, but why think that we should be *Darwinian* naturalists? One of Wittgenstein's better known comments is often cited as dismissive of naturalism, but this is not quite right. Wittgenstein says that 'Darwin's theory has no more to do with philosophy than any other hypothesis in natural science' (Wittgenstein 1961: 4.1122). Anyone who thought that all the sciences made roughly equal contributions to philosophy could agree with the letter of this claim, even if not with its spirit. What reasons are there, not merely to think that science is important to philosophy, but that Darwinian science is supremely important?

The philosopher Philip Kitcher usefully reminds Darwinian zealots that 'Anglo-American philosophers have explored a wide range of disciplines, using ideas from psychology, biology, political science, economics and the arts to reformulate traditional questions in epistemology and metaphysics' (Kitcher 1992: 55). We do not need to argue that Darwin's contribution to philosophy is pre-eminent in order to argue that it is important. We have seen ample evidence in this book of the fecundity of Darwin's views. In general there are two ways in which Darwinian ideas have helped us to find answers to philosophical problems (Flanagan 2003). On the one hand, the evolutionary sciences can provide a fund of

tools for thinking, or inspirations for analogy. Evolutionary epistemology, of the kind that sees scientific theories in competition
with each other, is just such an analogical extension of organic
biology. On the other hand, the evolutionary sciences can tell us
directly about the evolution of the human species and of various
human practices, including practices relating to our moral conduct,
our political behaviour, or the acquisition and use of knowledge.

We need to take some care when assessing claims about
Darwin's contribution to philosophy to clarify precisely what
'Darwinism', 'evolutionary theory' or 'the theory of natural selection', are supposed to be. Recall the controversy associated with
the value of the adaptive heuristic (see chapter five), which
recommends that we uncover the structure of the mind by reflection
on past environmental demands. It seems clear that if the adaptive
heuristic is to generate useful hypotheses regarding the probable
makeup of the human mind, then an abstract understanding of
natural selection must be supplemented with rich data from
anthropology, cognitive psychology, physiology, geology and
many other sciences. These disciplines together tell us which
adaptive problems we should recognise, and what solutions are
likely to be produced in response to them. If adaptive thinking
yields important answers to philosophical questions about, for
example, the nature of the emotions, then why claim a victory for
Darwinian naturalism, rather than for naturalism in general? After
all, many sciences will typically be involved. In a sense, the label
does not matter, so long as we remember that it is misleading to
suggest that there is one magical science – 'evolutionary biology' –
that gives us a special leg-up in doing philosophy; rather, the
effective practice of evolutionary biology demands that many
different branches of science cast their light on the problems we
are interested in.

Philip Kitcher captures this message when he argues in favour
of ' . . . bringing Darwin on to the philosophical team, not as the
star player who wins the day all by himself, but as a contributor to
a much larger effort' (Kitcher 2003: 400). Kitcher is right to
caution against a monomaniacal enthusiasm for the philosophical
promise of evolution, which shuts out other philosophically relevant

areas of learning. But Darwin himself – a great generalist, whose observations and readings drew from the fields of embryology, geology, economics, ethics, botany, animal behaviour and many others – is emblematic of this more eclectic conception of philosophical naturalism. Today, scientists are typically focused on extremely narrow fields of research. Philosophy has taken on the role of synthesising their results. Darwin's work reminds us of the importance of drawing together diverse natural and human sciences in order to provide a coherent picture of nature and our place in it.

SUMMARY

Darwin's work is important to mainstream philosophy because Darwin demonstrates that the human capacities which interest philosophers – the ability to know, to think, to praise and condemn – have histories, and that they can be studied in the ways we might study the capacities of any other species. This does not entail an abandonment of abstract philosophical theorising, but it does point the way towards closer integration between philosophy and the natural sciences. Darwin's work is also important in a broader philosophical sense. It changes how we think of ourselves. Darwin endorses a progressive conception of evolution, but Darwin's work does not show that our own species is the best the natural world has to offer, nor does it show that a species like ours is the sort of result that we should anticipate from the evolutionary process. Perhaps most important of all, we should resist the thought that a Darwinian view of human nature is primarily a biological view of human nature. Darwin's own work, and modern efforts to follow Darwin's lead, make social change part of the evolutionary process, and they make social organisation part of organic organisation.

FURTHER READING

Kitcher has written a sensible overview of Darwin's contribution to philosophical method:

Kitcher, P. (2003) 'Giving Darwin his Due', in J. Hodge and G. Radick (eds) *The Cambridge Companion to Darwin*, Cambridge: Cambridge University Press.

The dispute between Gould and Conway Morris regarding contingency is helpfully summarised in:

Sterelny, K. (2005) 'Another View of Life', *Studies in History and Philosophy of Biological and Biomedical Sciences*, 36: 585–93.

Darwin's views on progress, as well as the more recent dispute between Gould and Dawkins, are all covered in:

Shanahan, T. (2004) *The Evolution of Darwinism: Selection, Adaptation, and Progress in Evolutionary Biology*, Cambridge: Cambridge University Press.

Other useful works on progress and evolution include:

Gould, S. J. (1996) *Life's Grandeur: The Spread of Excellence from Plato to Darwin*, London: Penguin. (Published in the US under the title Full House.)
Sterelny, K. (2001b) *Dawkins vs. Gould: Survival of the Fittest*, Icon: Cambridge.

Major transition theory is introduced in:

Maynard Smith, J. and Szathmary, E. (2000) *The Origins of Life: From the Birth of Life to the Origins of Language*, Oxford: Oxford University Press.

Glossary

Adaptation Usually refers to any organic trait that is well-suited to its environment (such as an eye, or a wing), but it can also refer to the process by which such traits are generated. Contemporary biologists disagree about whether adaptation should be defined in terms of natural selection. Darwin sometimes uses the term 'co-adaptation' as a synonym.

Adaptive Heuristic The use of evolutionary problems encountered in a species' past to predict the likely traits selection has equipped species members with now. Sometimes called 'adaptive thinking'.

Affect Programs Emotions are understood as 'affect programs' by some modern psychologists. They are characterised as suites of responses (including expressive responses), triggered by particular kinds of stimuli.

Altruism A highly contested term. Behaviours are said to be biologically altruistic when they result in the fitness of the organism producing the behaviour being lower than the fitness of the organism that is the behaviour's beneficiary. Outside of biology, 'altruism' is frequently used in a psychological sense, to describe motivation that is in some sense selfless.

Argument from Design The argument that tries to show the existence of God on the basis of the good design in the natural world.

Artificial Selection The process by which animal breeders improve domesticated species.

Blending Inheritance Inheritance is said to be of a 'blending' form when the traits of parents blend together in their offspring. Genetic inheritance, by contrast, is said to be particulate, because discrete particles are transmitted from parents to offspring.

Catastrophism A view in geology which allows explanations of geological phenomena in terms of large-scale catastrophes of a kind not experienced by modern humans. It is contrasted with uniformitarianism, the view of Charles Lyell (among others). Lyell argues that geological explanation should only appeal to causes with which modern humans are familiar.

Epistemology The philosophical study of knowledge.

Fitness A concept not used by Darwin but one central to modern evolutionary biology. It is subject to many interpretations, but in its simplest forms refers to the ability of an organism to survive and reproduce.

Gemmules Particles which Darwin believed were responsible for inheritance.

Group Selection A process of natural selection that occurs between groups. Darwin usually writes in terms of selection at the level of the 'community', rather than the group. Modern biologists are divided on how to understand what group selection is, and on whether it is an important evolutionary process.

Higher Taxa Units of biological classification at levels higher than the species. Examples might include kingdoms, phyla or classes.

Individuals Entities with a beginning and end in time, and spatial boundaries. Some biologists and philosophers think that species are individuals, for they begin when the species is first formed, they end when the species becomes extinct, and their physical boundaries coincide with the area occupied by the organisms within the species. This is contrasted with the view of species as kinds.

Inference to the Best Explanation A slogan in the general philosophy of science which captures the thought that we should believe the theory which best explains some set of phenomena. Darwin frequently argues that the explanatory power of his evolutionary views (compared with the views of special creationists) means that his views are more likely to be true.

Inheritance of Acquired Characteristics A process (now discredited) by which traits acquired by an organism during the course of its life appear in offspring. The standard example is of the blacksmith, whose arm gets stronger through exercise, and whose sons are born with strong arms as a result of this. Darwin was a believer in the inheritance of acquired characteristics.

Intelligent Design Theory A modern day version of the design argument, popular among non-scientists in the United States. The theory says that organic life bears the marks of intelligent design, although it declines to comment on the characteristics of the designer, including the designer's divinity.

Interactor In modern formulations of evolutionary theory, this is the role played by any entity in a natural selection process that interacts with an environment in such a way as to result in the differential production of replicators.

Kinds Sets of resembling objects. Some philosophers argue that species are kinds – i.e. sets of resembling organisms – and some of Darwin's remarks suggest he has sympathies with this view. This conception of species is usually contrasted with the view of species as individuals.

Lamarckism The view of evolution associated with the pre-Darwinian biologist Lamarck. Lamarckism is often contrasted with Darwinism, and Lamarck himself is often derided for his belief in the importance of the inheritance of acquired characteristics. But Darwin, too, believed in this mode of inheritance. Lamarck's views differed from Darwin's in many other ways: Lamarck thought that individual organisms had an inherent tendency to adapt to their surroundings, and although he believed in the unlimited modifiability of species, he did not subscribe to Darwin's image of the Tree of Life.

Likelihood A concept used in statistics and in philosophical theories of evidence. The likelihood of a hypothesis H in the light of some evidence E is the probability of E given H. This is not to be confused with the probability of H given E. Suppose we observe that Tony Blair has a metabolism. This does not make it probable that he is a Martian. Even so, the hypothesis that Tony Blair is a Martian makes it probable that Tony Blair has a metabolism. Hence the hypothesis that Tony Blair is a Martian has high likelihood in light of the observation that he has a metabolism.

Meme In theories of cultural evolution, memes are units analogous to genes. Most memeticists think of ideas as memes; other candidate memes might include techniques, tunes, and for some theorists tools. Memes, like genes, are a type of replicator.

Meta-ethics The study of moral discourse and its subject matter. Questions in meta-ethics include such things as whether there are moral facts, whether moral judgements are expressions of emotion or claims about states of the world, and so forth. Meta-ethics is distinguished from normative ethics, the study of what should be the case. Questions in normative ethics might include, 'Should we permit abortion?' or 'How much weight should we give to the interests of future generations when planning today?'.

Metaphysics In modern philosophy, metaphysics refers to the study of the basic nature of the universe. Questions in metaphysics might include 'What is causation?' or 'What is the difference between past, present and future?'. In Darwin's time, metaphysics instead referred specifically to issues about the mind.

Modern Synthesis The form of evolutionary biology constructed in the 1920s and 1930s, which combined a Darwinian belief in the importance of natural selection with a genetic theory of inheritance. Today's evolutionary biology is very similar to that of the modern synthesis.

Modularity To say that the mind is modular is to say that it is composed of several specialised information-processing tools or 'modules'. In evolutionary psychology, the Santa Barbara School argues that the mind is 'massively modular', with hundreds or even thousands of modules, all built to deal with different environmental problems. The precise characterisation of modules is contentious, but the Santa Barbara School also claims that they are innate. The notion of modularity is also invoked in developmental biology, where it means something quite different. Developmental modules are physiological units whose development is subject to (largely) independent control.

Naturalism In philosophy, naturalism is the view that philosophical work on, for example, the mind, or knowledge, should be informed by work in the natural sciences, especially psychological and biological sciences. There are many forms of naturalism, some of which argue that all the phenomena that can be explained should be explained in scientific terms.

Natural Kinds Chemical elements are the canonical examples of natural kinds. They are basic types of stuff, which science seeks to identify and to characterise. Natural kinds are usually understood to exist independently of scientific investigators and their interests. Although philosophers often place biological species in their standard lists of natural kinds, those biologists who argue that species are individuals oppose this.

Natural Selection The principle devised by Darwin to account for the adaptation of organisms to their environments. Darwin also invoked natural selection to explain the generation of new species. Modern biologists tend to say that natural selection occurs whenever there is 'heritable variation in fitness'; roughly, that means natural selection occurs whenever organisms in the parental generation vary in their abilities to survive and reproduce, so long as offspring resemble their parents. Darwin understands natural selection in a way that is tailored more directly to the explanation of adaptation.

Natural Theology A movement in natural history that was especially influential in Britain in the seventeenth and eighteenth centuries, natural theology sought to identify the character of God based on observations of organic nature.

Niche-construction The process by which organisms do not merely adapt passively to a stable environment, but actively maintain and alter that environment.

Normative Ethics See Meta-ethics.

Pangenesis Darwin's theory of inheritance. He believed that sperm and egg cells each contained particles called gemmules, which were originally produced by each of the organs of the body, and which migrated to the sex cells. Gemmules in an embryo could either lie dormant, or develop to produce organs which resembled the parental organs that had produced them.

Particulate Inheritance See Blending Inheritance.

Pleistocene The period of time from about 1.8 million years ago to 10,000 years ago, during which time our species is generally believed to have led a hunter-gatherer existence. This period is important to evolutionary psychologists of the Santa Barbara

School, for they believe that this was the period when our characteristic cognitive adaptations were formed.

Population Thinking According to Ernst Mayr, population thinking was Darwin's third great conceptual innovation, after natural selection and the Tree of Life hypothesis. There are a number of themes associated with population thinking, all of which are set up in contrast with typological thinking. In broad terms, the typologist believes there are only a few stable forms underlying organic variation; the population thinker denies this.

Principle of Divergence of Character For Darwin, this was an important principle in the explanation of speciation. Darwin believed that a uniform species could split into two or more distinct species, because a collection of distinct specialised forms would be better equipped than a uniform group of generalists to take advantage of the many different opportunities available in a single physical environment.

Replicator In modern presentations of evolutionary theory, replicators are units that are able to make copies of themselves, thereby ensuring that offspring generations resemble parents. Genes are usually taken to be canonical replicators, but some argue that there are others, such as memes.

Reverse Engineering The investigative process whereby one attempts to discern the environmental problems that were responsible for some organic behaviour or characteristic.

Sexual Selection The process by which the struggle to find a mate leads to behavioural, anatomical or psychological modification. Darwin believed that sexual selection was important throughout the animal kingdom. He thought it could account for the gaudy plumage of male birds, and he also believed it accounted for differences between human races and human sexes.

Special Creation In Darwin's time, the name for the view that each species was individually created by supernatural influence.

Speciation The processes by which new species are formed.

Struggle for Existence A key element in Darwin's presentation of natural selection, the struggle for existence occurs as a result of population growth outstripping the food supply available to a species. This in turn leads to the preferential survival of those

organisms best suited to the local conditions. Darwin sometimes uses the alternative phrase 'Struggle for Life'. He is
explicit that he means 'struggle' to be read in a metaphorical
way, but some modern synthesis biologists, especially R. A.
Fisher, deny that any form of struggle is essential to natural
selection.

Subversion from Within A problem for the efficacy of group
selection. Even if one group is fitter than another by virtue of
possessing a large number of altruistic organisms, any selfish
organisms within the group are likely to be fitter than the altruists, and they are likely to out-compete them. Hence group
selection is undermined as a mechanism to explain the evolution of altruism by the threat that altruistic groups will be
subverted from within by their own selfish members.

Taxonomy In general, taxonomy is the study and practice of
sorting items of any kind into classes. There can be a taxonomy
of library books, which focuses on how best to arrange books
under different subject headings. Biological taxonomy is the
study and practice of assigning organisms to various categories,
including species, families and phyla.

Teleology A term in philosophy, referring to the study of ends,
goals, and goal-directed systems. Attempts to understand the
sense in which eyes can be said to be 'for' seeing fall within the
domain of teleology.

Transformisme The French name for transmutationism, or
evolution.

Transmutationism In Darwin's time, the view that species were
not fixed, but might instead be transformed over time, or that
one species might split into two or more new and distinct
species.

Tree of Life Darwin claims that life can be considered to form a
great tree, which depicts the genealogical relations between
species. When modern biologists say that life has evolved, they
are saying that species are related in such a way that they form
a tree-like structure of descent.

Typological Thinking See Population Thinking.

Uniformitarianism See Catastrophism.

Use-Inheritance The common British name for the process, accepted by Darwin and most of his contemporaries, whereby an organ that is habitually used in a certain way during the life of an individual organism is inherited in strengthened form in the organism's offspring. Darwin also believed that a behaviour habitually performed would be inherited in the offspring as automated instinct. See also Inheritance of Acquired Characteristics.

Utilitarianism The view in ethics that the right action is that which produces the best overall consequences for human welfare. In its classic form, utilitarianism says that the right action is that which produces the greatest happiness of the greatest number.

Variety A single species can be composed of many different varieties, usually distinguished on the basis of appearance or behaviour. Individuals of different varieties can mate successfully with each other, unlike individuals of different species.

Vera Causa A 'true cause'. Darwin's contemporaries argued that only some explanatory causes were scientifically legitimate. Others were wholly speculative, and not to be trusted. The trick was to come up with plausible methodological principles that would say which causes were *bona fide*, and which were speculative. John Herschel's way of picking out *verae causae* greatly influenced Darwin.

References

Allen, C., Bekoff, M. and Lauder, G. (1998) *Nature's Purposes: Analyses of Function and Design in Biology*, Cambridge, MA: MIT Press.

Amundson, R. (2005) *The Changing Role of the Embryo in Evolutionary Thought*, Cambridge: Cambridge University Press.

Ariew, A. (1999) 'Innateness is Canalization', in V. Hardcastle (ed.) *Where Biology Meets Psychology*, Cambridge, MA: MIT Press.

Aunger, R. (2000) 'Introduction', in R. Aunger (ed.) *Darwinizing Culture*, Oxford: Oxford University Press.

Bacon, F. (2000) *The Advancement of Learning*, Oxford: Clarendon Press; originally published 1605.

Barkow, J., Cosmides, L. and Tooby, J. (1992) *The Adapted Mind: Evolutionary Psychology and the Generation of Culture*, Oxford: Oxford University Press.

Barrett, P., Gautrey, P., Herbert, S., Kohn, D. and Smith, S. (1987) *Charles Darwin's Notebooks, 1836–1844*, Cambridge: Cambridge University Press.

Behe, M. (1996) *Darwin's Black Box: The Biochemical Challenge to Evolution*, New York: Simon and Schuster.

Binmore, K. (2005) *Natural Justice*, Oxford: Oxford University Press.

Bowler, P. (1984) *Evolution: The History of an Idea*, Los Angeles: University of California Press.

Bowler, P. (1990) *Charles Darwin*, Cambridge: Cambridge University Press.

Boyd, R. (1991) 'Realism, Anti-foundationalism, and the Enthusiasm for Natural Kinds', *Philosophical Studies*, 61: 127–48.

Boyd, R. and Richerson, P. (2000) 'Memes: Universal Acid or a Better Mousetrap?', in R. Aunger (ed.) *Darwinizing Culture*, Oxford: Oxford University Press.

Brooke, J. (2003) 'Darwin and Victorian Christianity', in J. Hodge and G. Radick (eds) *The Cambridge Companion to Darwin*, Cambridge: Cambridge University Press.

Browne, J. (2003a) *Charles Darwin: Voyaging*, London: Pimlico.

Browne, J. (2003b) *Charles Darwin: The Power of Place*, London: Pimlico.

Buller, D. (2005) *Adapting Minds: Evolutionary Psychology and the Persistent Quest for Human Nature*, Cambridge, MA: MIT Press.

Buss, D. (1989) 'Sex Differences in Human Mate Preferences: Evolutionary Hypotheses Tested in 37 Cultures', *Behavioral and Brain Sciences*, 12: 1–14.

Buss, D. (1994) *The Evolution of Desire: Strategies of Human Mating*, New York: Basic Books.

Buss, D. (1999) *Evolutionary Psychology: The New Science of the Mind*, London: Allyn and Bacon.

Buss, L. (1987) *The Evolution of Individuality*, Princeton, NJ: Princeton University Press.

Campbell, D. T. (1974) 'Evolutionary Epistemology', in P. Schilpp (ed.) *The Philosophy of Karl Popper*, LaSalle, IL: Open Court.

Charnov, E. and Bull, J. (1977) 'When is Sex Environmentally Determined?', *Nature*, 266: 828–30.

Chevalier-Skolnikoff, S. (1973) 'Facial Expression of Emotion in Nonhuman Primates', in P. Ekman (ed.) *Darwin and Facial Expression: A Century of Research in Review*, New York: Academic Press.

Conway Morris, S. (2003) *Life's Solution: Inevitable Humans in a Lonely Universe*, Cambridge: Cambridge University Press.

Cosmides, L and Tooby, J. (1997a) 'The Modular Nature of Human Intelligence', in A. B. Scheibel and J. W. Schopf (eds) *The Origin and Evolution of Intelligence*, Sudbury, MA: Jones and Bartlett.

Cosmides, L. and Tooby, J. (1997b) 'Letter to the Editor of *The New York Review of Books* on Stephen Jay Gould's "Darwinian Fundamentalism"' (June 12, 1997) and "Evolution: The Pleasures of Pluralism" (June 26, 1997)' Online. Available: http://cogweb.ucla.edu/Debate/CEP_Gould.html (accessed 26th November 2005).

Cosmides, L., Tooby, J. and Barkow, J. (1992) 'Introduction', in J. Barkow, L. Cosmides and J. Tooby (eds) *The Adapted Mind: Evolutionary Psychology and the Generation of Culture*, Oxford: Oxford University Press.

Coyne, J. (2000) 'The Fairy Tales of Evolutionary Psychology: Of Vice and Men', *The New Republic*, 3 April: 27–34.

Coyne, J. and Orr, H. A. (2004) *Speciation*, Sunderland, MA: Sinauer Associates.

Cronin, H. (2004) 'Getting Human Nature Right', in J. Brockman (ed.) *Science at the Edge*, London: Weidenfeld and Nicholson.

Daeschler, E., Shubin, N. and Jenkins, F. (2006) 'A Devonian Tetrapod-like Fish and the Evolution of the Tetrapod Body Plan', *Nature*, 440: 757–63.

Darwin, C. (1903) *More Letter of Charles Darwin, Volume II*, F. Darwin and A. Seward (eds) London: John Murray.

Darwin, C. (1905) *The Life and Letters of Charles Darwin, Volume II*, F. Darwin (ed.) New York: D. Appleton and Co; first published by John Murray (1887).

Darwin, C. (1913) *Journal of Researches*, eleventh edition; first published by John Murray (1839).

Darwin, C. (1959) *The Origin of Species by Charles Darwin: A Variorum Text*, M. Peckham (ed.) Philadelphia: University of Pennsylvania Press.

Darwin, C. (1989) *The Voyage of the* Beagle, J. Browne and M. Neve (eds) London: Penguin Classics.

Darwin, C. (1998) *The Expression of the Emotions in Man and Animals*, London: HarperCollins, first published in 1889, third edition.

Darwin, C. (2002) *Autobiographies*, M. Neve and S. Messenger (eds) London: Penguin Classics.

Darwin, C. (2004) *The Descent of Man*, London: Penguin Classics, first published 1877, second edition.

Dawkins, R. (1976) *The Selfish Gene*, Oxford: Oxford University Press.

Dawkins, R. (1986) *The Blind Watchmaker*, London: Penguin.

Dawkins, R. (1997) 'Human Chauvinism', *Evolution*, 51: 1015–20.

Dembski, W. (1998) *The Design Inference: Eliminating Chance Through Small Probabilities*, Cambridge: Cambridge University Press.

Dembski, W. (2001) 'Intelligent Design as a Theory of Information', in R. Pennock (ed.), *Intelligent Design Creationism and its Critics*, Cambridge, MA: MIT Press.

Dembski, W. (2004) 'The Logical Underpinnings of Intelligent Design', in W. Dembski and M. Ruse (eds) *Debating Design: From Darwin to DNA*, Cambridge: Cambridge University Press.

Dennett, D. C. (1995) *Darwin's Dangerous Idea: Evolution and the Meanings of Life*, New York: Norton.

Dennett, D. C. (2006) *Breaking the Spell: Religion as a Natural Phenomenon*, London: Allen Lane.

Depew, D. and Weber, B. (1996) *Darwinism Evolving: Systems Dynamics and the Genealogy of Natural Selection*, Cambridge, MA: MIT Press.

Desmond, A. and Moore, J. (1992) *Darwin*, London: Penguin.

Diamond, J. (1996) *Guns, Germs and Steel*, New York: Norton.

Dobzhansky, T. (1951) *Genetics and the Origin of Species*, third edition, New York: Columbia University Press.

Dupré, J. (2002) *Humans and Other Animals*, Oxford: Oxford University Press.

Ekman, P. (1973) 'Cross-Cultural Studies of Facial Expression', in P. Ekman (ed.) *Darwin and Facial Expression: A Century of Research in Review*, New York: Academic Press.

Endersby, J. (2003) 'Darwin on Generation, Pangenesis and Sexual Selection', in J. Hodge and G. Radick (eds) *The Cambridge Companion to Darwin*, Cambridge: Cambridge University Press.

Ereshefsky, M. and Matthen, M. (2005) 'Taxonomy, Polymorphism and History: An Introduction to Population Structure Theory', *Philosophy of Science*, 72: 1–21.

Flanagan, O. (2003) 'Ethical Expressions: Why Moralists Scowl, Frown and Smile', in J. Hodge and G. Radick (eds) *The Cambridge Companion to Darwin*, Cambridge: Cambridge University Press.

Gayon, J. (1998) *Darwinism's Struggle for Survival*, Cambridge: Cambridge University Press.

Geertz, C. (1973) *The Interpretation of Cultures: Selected Essays*, New York: Basic Books.

Ghiselin, M. (1969) *The Triumph of the Darwinian Method*, Berkeley, CA: University of California Press.

Ghiselin, M. (1973) 'Darwin and Evolutionary Psychology', *Science*, 179: 964–68.

Ghiselin, M. (1974) 'A Radical Solution to the Species Problem', *Systematic Zoology*, 23: 536–44.

Ghiselin, M. (1994) 'Darwin's Language may Seem Teleological, but his Thinking is Another Matter', *Biology and Philosophy*, 9: 489–92.

Godfrey-Smith, P. (1996) *Complexity and the Function of Mind in Nature*, Cambridge: Cambridge University Press.

Godfrey-Smith, P. (2000) 'The Replicator in Retrospect', *Biology and Philosophy*, 15: 403–23.

Gould, S. J. (1988) 'On Replacing the Idea of Progress with an Operational Notion of Directionality', in M. Nitecki (ed.) *Evolutionary Progress*, Chicago: Chicago University Press.

Gould, S. J. (1991) *Wonderful Life: The Burgess Shale and the Nature of History*, London: Penguin.

Gould, S. J. (1996) *Life's Grandeur: The Spread of Excellence from Plato to Darwin*, London: Penguin; published in the United States under the title *Full House*.

Gould, S. J. and Lewontin, R. (1979) 'The Spandrels of San Marco and the Panglossian Paradigm: A Critique of the Adaptationist Programme', *Proceedings of the Royal Society*, B205: 581–98.

Gray, R., Heaney, M. and Fairhall, S. (2003) 'Evolutionary Psychology and the Challenge of Adaptive Explanation', in K. Sterelny and J. Fitness (eds) *From Mating to Mentality: Evaluating Evolutionary Psychology*, London: Psychology Press.

Griffiths, P. (1997) *What Emotions Really Are: The Problem of Psychological Categories*, Chicago: University of Chicago Press.

Griffiths, P. (1999) 'Squaring the Circle: Natural Kinds with Historical Essences', in R. Wilson (ed.) *Species: New Interdisciplinary Essays*, Cambridge, MA: MIT Press.

Griffiths, P. (2002) 'What is Innateness?', *Monist*, 85: 70–85.

Griffiths, P. and Gray, R. (1994) 'Developmental Systems and Evolutionary Explanation', *Journal of Philosophy*, 91: 277–304.

Hacking, I. (1983) *Representing and Intervening*, Cambridge: Cambridge University Press.

Herschel, J. (1996) *A Preliminary Discourse on the Study of Natural Philosophy*, London: Routledge/Thoemmes Press; reprint of the 1830 edition.

Hesse, M. (1966) *Models and Analogies in Science*, Notre Dame, IN: University of Notre Dame Press.

Hodge, J. and Radick, G. (eds) (2003) *The Cambridge Companion to Darwin*, Cambridge: Cambridge University Press.

Hodge, M. J. S. (1977) 'The Structure and Strategy of Darwin's "Long Argument"', *British Journal for the History of Science*, 10: 237–46.

Hodge, M. J. S. (2000) 'Knowing about Evolution: Darwin and his Theory of Natural Selection', in R. Creath and J. Maienshein (eds) *Biology and Epistemology*, Cambridge: Cambridge University Press.

Hodge, M. J. S. (2003) 'The Notebook Programmes and Projects of Darwin's London Years', in J. Hodge and G. Radick (eds) *The Cambridge Companion to Darwin*, Cambridge: Cambridge University Press.

Hull, D. (1978) 'A Matter of Individuality', *Philosophy of Science*, 45: 335–60.

Hull, D. (1988) *Science as a Process*, Chicago: University of Chicago Press.

Hull, D. (1998) 'On Human Nature', in D. Hull and M. Ruse (eds) *The Philosophy of Biology*, Oxford: Oxford University Press; originally published in A. Fine and P. Machamer (eds) *PSA Volume Two* (1986): 3–13.

Hull, D. (2001) 'A Mechanism and its Metaphysics: An Evolutionary Account of the Social and Conceptual Development of Science', in D. Hull, *Science and Selection*, Cambridge: Cambridge University Press.

James, W. (1880) 'Great Men, Great Thoughts, and the Environment', *Atlantic Monthly*, 66: 441–59.

Johnson, T., Scholz, C., Talbot, M., Kelts, K., Ricketts, R., Ngobi, G., Beuning, K., Ssemmanda, I. and McGill, J. (1996) 'Late Pleistocene Desiccation of Lake Victoria and Rapid Evolution of Cichlid Fishes', *Science*, 273: 1091–93.

Kenrick, D. and Keefe, R. (1992) 'Age Preferences in Mates Reflect Sex Differences in Reproductive Strategies', *Behavioral and Brain Sciences*, 15: 75–91.

Kevles, D. (1995) *In the Name of Eugenics*, Cambridge, MA: Harvard University Press.

Kingsolver, J. and Koehl, M. (1985) 'Aerodynamics, Thermoregulation, and the Evolution of Insect Wings: Differential Scaling and Evolutionary Change', *Evolution*, 39: 488–504.

Kitcher, P. (1992) 'The Naturalists Return', *Philosophical Review*, 101: 53–114.

Kitcher, P. (1994) 'Four Ways of "Biologicizing" Ethics', in E. Sober (ed.) *Conceptual Issues in Evolutionary Biology*, second edition, Cambridge, MA: MIT Press.

Kitcher, P. (2003) 'Giving Darwin his Due', in J. Hodge and G. Radick (eds) *The Cambridge Companion to Darwin*, Cambridge: Cambridge University Press.

Krebs, J. and Davies, N. (1997) 'Introduction to Part Two', in J. Krebs and N. Davies (eds) *Behavioural Ecology: An Evolutionary Approach*, fourth edition, Oxford: Blackwell Science.

Kuper, A. (2000) 'If Memes are the Answer, What is the Question?', in R. Aunger (ed.) *Darwinizing Culture*, Oxford: Oxford University Press.

Laland, K. and Brown, G. (2002) *Sense and Nonsense: Evolutionary Perspectives on Human Behaviour*, Oxford: Oxford University Press.

Laudan, L. (1981) 'William Whewell on the Consilience of Inductions', in L. Laudan, *Science and Hypothesis: Historical Essays on Scientific Methodology*, Reidel: Dordrecht.

Lewens, T. (2004) *Organisms and Artifacts: Design in Nature and Elsewhere*, Cambridge, MA: MIT Press.

Lewens, T. (2005) 'The Problems of Biological Design', in A. O'Hear (ed.) *Philosophy, Biology and Life*, Cambridge: Cambridge University Press.

Lewontin, R. C. (1978) 'Adaptation', *Scientific American*, 239(3): 212–30.

Lewontin, R. C. (1985) 'The Analysis of Variance and the Analysis of Causes', in R. Levins and R. Lewontin (eds) *The Dialectical Biologist*, Cambridge, MA: Harvard University Press; originally published in *American Journal of Human Genetics*, 26 (1974): 400–11.

Lipton, P. (2004) *Inference to the Best Explanation*, second edition. London: Routledge.

Lutz, C. (1988) *Unnatural Emotions: Everyday Sentiments on a Micronesian Atoll and their Challenge to Western Theory*, Chicago: University of Chicago Press.

Maguire, E., Gadian, D., Johnsrude, I., Good, C., Ashburner, J., Frackowiak, R. and Frith, C. (2000) 'Navigation-related Structural Change in the Hippocampi of Taxi Drivers', *PNAS*, 97: 4398–403.

Mallon, R. and Stich, S. (2000) 'The Odd Couple: The Compatibility of Social Construction and Evolutionary Psychology', *Philosophy of Science*, 67: 133–54.

Mameli, M. (2004) 'Nongenetic Selection and Nongenetic Inheritance', *British Journal for the Philosophy of Science*, 55: 35–71.

Mameli, M. and Bateson, P. (2006) 'Innateness and the Sciences', *Biology and Philosophy*, 21: 155–88.

Marx, K. and Engels, F. (1936) *Karl Marx and Friedrich Engels. Correspondence, 1846–1895: A Selection with Commentary and Notes*, trans. D. Torr, London: M. Lawrence.

Maynard Smith, J. and Parker, G. A. (1976) 'The Logic of Asymmetric Contests', *Animal Behaviour*, 24: 159–75.

Maynard Smith, J. and Szathmary, E. (1995) *The Major Transitions in Evolution*, Oxford: W. H. Freeman/Spektrum.

Maynard Smith, J. and Szathmary, E. (2000) *The Origins of Life: From the Birth of Life to the Origins of Language*, Oxford: Oxford University Press.

Mayr, E. (1976) 'Typological Versus Population Thinking', in E. Mayr, *Evolution and the Diversity of Life*, Cambridge, MA: Harvard University Press.

Mellor, D. H. (1976) 'Probable Explanation', *Australasian Journal of Philosophy*, 54: 231–41.

Miller, G. (2000) *The Mating Mind: How Sexual Choice Shaped the Evolution of Human Nature*, London: Heineman.

Mivart, St G. (1871) *The Genesis of Species*, second edition, London: Macmillan.

Moore, G. E. (1903) *Principia Ethica*, Cambridge: Cambridge University Press.

Neander, K. (1995) 'Pruning the Tree of Life', *British Journal for the Philosophy of Science*, 46: 59–80.

Nietzsche, F. (1974) *The Gay Science*, trans. Walter Kaufmann, New York: Random House; first published 1881.

Okasha, S. (2001) 'Why Won't the Group Selection Controversy Go Away?', *British Journal for the Philosophy of Science*, 52: 25–50.

Paquin, C. and Adams, J. (1983) 'Relative Fitness Can Decrease in Evolving Asexual Populations of S. *cerevisiae*', *Nature*, 306: 368–71.

Paul, D. (2003) 'Darwin, Social Darwinism, and Eugenics', in J. Hodge and G. Radick (eds) *The Cambridge Companion to Darwin*, Cambridge: Cambridge University Press.

Paul, D. and Spencer, H. (2001) 'Did Eugenics Rest on an Elementary Mistake?', in R. Singh, C. Krimbas, D. Paul and J. Beatty (eds) *Thinking About Evolution*, Cambridge: Cambridge University Press.

Pinker, S. (1997) *How the Mind Works*, New York: Norton.

Popper, K. (1935) *Logik der Forschung*, Vienna: Springer, translated as *The Logic of Scientific Discovery*, London: Hutchinson, 1959.

Popper, K. (1962) 'Conjectures and Refutations', in K. Popper, *Conjectures and Refutations*, Routledge: London.

Provine, W. (1971) *The Origins of Theoretical Population Genetics*, Chicago: Chicago University Press.

Railton, P. (1986) 'Moral Realism', *Philosophical Review*, 95: 163–207.

Rehbock, P. (1983) *The Philosophical Naturalists: Themes in Early Nineteenth-Century British Biology*, Madison: University of Wisconsin Press.

Richards, E. (1983) 'Darwin and the Descent of Women', in D. Oldroyd and I. Langham (eds) *The Wider Domain of Evolutionary Thought*, Dordrecht: Reidel.

Richards, R. (1983) 'Why Darwin Delayed', *Journal of the History of the Behavioral Sciences*, 19: 45–53.

Richards, R. (1987) *Darwin and the Emergence of Evolutionary Theories of Mind and Behavior*, Chicago: University of Chicago Press.

Richerson, P. and Boyd, R. (2005) *Not by Genes Alone: How Culture Transformed Human Evolution*, Chicago: University of Chicago Press.

Ridley, M. (1996) *Evolution*, second edition, Oxford: Blackwell Science.

Rosenberg, A. (2003) 'Darwinism in Moral Philosophy and Social Theory', in J. Hodge and G. Radick (eds) *The Cambridge Companion to Darwin*, Cambridge: Cambridge University Press.

Ruse, M. (1975) 'Darwin's Debt to Philosophy: An Examination of the Influence of the Philosophical Ideas of John F. W. Herschel and William Whewell on the Development of Charles Darwin's Theory of Evolution', *Studies in History and Philosophy of Science*, 6: 159–81.

Ruse, M. (2000a) 'Darwin and the Philosophers: Epistemological Factors in the Development and Reception of the Theory of the *Origin of Species*', in R. Creath and J. Maienshein (eds) *Biology and Epistemology*, Cambridge: Cambridge University Press.

Ruse, M. (2000b) *Can a Darwinian be a Christian? The Relationship between Science and Religion*, Cambridge: Cambridge University Press.

Ruse, M. and Wilson, E. O. (1993) 'Moral Philosophy as Applied Science' reprinted in Sober, E. (ed.) *Conceptual Issues in Evolutionary Biology*, second edition, Cambridge, MA: MIT Press; first published in *Philosophy* 61 (1986): 173–92.

Russell, J. (1991) 'Culture and the Categorization of Emotions', *Psychological Bulletin*, 110: 426–50.

Segerstråle, U. (2000) *Defenders of the Truth*, Oxford: Oxford University Press.

Shanahan, T. (2004) *The Evolution of Darwinism: Selection, Adaptation, and Progress in Evolutionary Biology*, Cambridge: Cambridge University Press.

Skyrms, B. (1996) *Evolution of the Social Contract*, Cambridge: Cambridge University Press.

Shuster, S. and Wade, M. (1991) 'Equal Mating Success among Male Reproductive Strategies in a Marine Isopod', *Nature*, 350: 608–10.

Singer, P. (1999) *A Darwinian Left: Politics, Evolution and Cooperation*, London: Weidenfeld and Nicolson.

Sloan, P. (2003) 'The Making of a Philosophical Naturalist', in J. Hodge and G. Radick (eds) *The Cambridge Companion to Darwin*, Cambridge: Cambridge University Press.

Sober, E. (1980) 'Evolution, Population Thinking, and Essentialism', *Philosophy of Science*, 47: 350–83.

Sober, E. (1984) *The Nature of Selection: Evolutionary Theory in Philosophical Focus*, Cambridge, MA: MIT Press.

Sober, E. (1993) *Philosophy of Biology*, Boulder, CO: Westview.

Sober, E. (1994a) 'Prospects for an Evolutionary Ethics', in E. Sober, *From a Biological Point of View*, Cambridge: Cambridge University Press.

Sober, E. (1994b) 'Did Evolution make us Psychological Egoists?', in E. Sober, *From a Biological Point of View*, Cambridge: Cambridge University Press.

Sober, E. (1994c) 'The Adaptive Advantage of Learning and *A Priori* Prejudice', in E. Sober, *From a Biological Point of View*, Cambridge: Cambridge University Press.

Sober, E. (2004) 'The Design Argument', in W. Demski and M. Ruse (eds) *Debating Design: From Darwin to DNA*, Cambridge: Cambridge University Press.

Sober, E. and Wilson, D. S. (1998) *Unto Others: The Evolution and Psychology of Unselfish Behavior*, Cambridge, MA: Harvard University Press.

Spencer, H. (1855) *Principles of Psychology*, London: Longman, Brown, Green and Longmans.

Sperber, D. (2000) 'An Objection to the Memetic Approach to Culture', in R. Aunger (ed.) *Darwinizing Culture*, Oxford: Oxford University Press.

Sterelny, K. (2001a) 'Niche Construction, Developmental Systems and the Extended Replicator', in R. Gray, P. Griffiths and S. Oyama (eds) *Cycles of Contingency*, Cambridge, MA: MIT Press.

Sterelny, K. (2001b) *Dawkins vs. Gould: Survival of the Fittest*, Cambridge: Icon.

Sterelny, K. (2003a) *Thought in a Hostile World: The Evolution of Human Cognition*, Oxford: Blackwell.

Sterelny, K. (2003b) 'Darwinian Concepts in the Philosophy of Mind', in J. Hodge and G. Radick (eds) *The Cambridge Companion to Darwin*, Cambridge: Cambridge University Press.

Sterelny, K. (2005) 'Another View of Life', *Studies in History and Philosophy of Biological and Biomedical Sciences*, 36: 585–93.

Sterelny, K. and Griffiths, P. (1999) *Sex and Death: An Introduction to the Philosophy of Biology*, Chicago: University of Chicago Press.

Sterelny, K., Smith, K. and Dickison, M. (1996) 'The Extended Replicator', *Biology and Philosophy*, 11: 377–403.

Sulloway, F. (1982) 'Charles Darwin's Finches: The Evolution of a Legend', *Journal of the History of Biology*, 15: 1–53.

Sunstein, C. (2005) *The Laws of Fear*, Cambridge: Cambridge University Press.

Thornhill, R. and Palmer, C. (2000) *A Natural History of Rape: Biological Bases of Sexual Coercion*, Cambridge, MA: MIT Press.

Tooby, J. and Cosmides, L. (1990) 'On the Universality of Human Nature and the Uniqueness of the Individual: The Role of Genetics and Adaptation', *Journal of Personality*, 58: 17–67.

Tooby, J. and Cosmides, L. (1992) 'The Psychological Foundations of Culture', in J. Barkow, L. Cosmides and J. Tooby (eds) *The Adapted Mind: Evolutionary Psychology and the Generation of Culture*, Oxford: Oxford University Press.

Townsend, J. and Levy, G. (1990) 'Effects of Potential Partners' Costume and Physical Attractiveness on Sexuality and Partner Selection', *Journal of Psychology*, 124: 371–89.

Warner, D. and Shine, R. (2005) 'The Adaptive Significance of Temperature-Dependent Sex Determination: Experimental Tests with a Short-lived Lizard', *Evolution*, 59: 2209–21.

Waters, K. (2003) 'The Arguments in the Origin of Species', in J. Hodge and G. Radick (eds) *The Cambridge Companion to Darwin*, Cambridge: Cambridge University Press.

Whewell, W. (1833) *Astronomy and General Physics Considered with Reference to Natural Theology*, London: William Pickering.

Whewell, W. (1996) *The Philosophy of the Inductive Sciences: Volume II*, London: Routledge/Thoemmes Press; reprint of the 1840 edition.

Williams, G. C. (1966) *Adaptation and Natural Selection*, Princeton: Princeton University Press.

Williams, G. C. (1996) *Plan and Purpose in Nature*, London: Weidenfeld and Nicholson.

Wilson, D. S. (1994) 'Adaptive Genetic Variation and Evolutionary Psychology', *Ethology and Sociobiology*, 15: 219–35.

Wilson, D. S. (2002) *Darwin's Cathedral: Evolution, Religion and the Nature of Society*, Chicago: University of Chicago Press.

Wilson, E. O. (1975) *Sociobiology: The New Synthesis*, Cambridge, MA: Harvard University Press.

Wittgenstein, L. (1961) *Tractatus Logico-Philosophicus*, London: Routledge.

Young, M. and Edis, T. (eds) (2004) *Why Intelligent Design Fails: A Scientific Critique of the New Creationism*, New Brunswick: Rutgers University Press.

Index

RELATED TITLES FROM ROUTLEDGE

HUMAN NATURE AFTER DARWIN
Janet Radcliffe Richards

'Janet Radcliffe Richards reveals the real "implications" of Darwinism for our view of ourselves. If you knew that the anti-Darwinians must be wrong but you lacked ammunition – here it is.' – *Helena Cronin, London School of Economics*

'Janet Radcliffe Richards has scored yet another success. Human Nature After Darwin is simply the clearest and most accurate introduction that there is to the current controversies about evolution, about Darwinian evolution in particular, and about how these do or do not apply to our own species. This is a book that will prove invaluable to students of all ages. Highly recommended.' – *Michael Ruse, University of Guelph, Canada*

'A really excellent text. Richards uses the controversy over sociobiology as a way to discuss a whole series of traditional philosophical problems. This is an introductory text dealing with extremely important issues.' – *David Hull, Northwestern University*

'This book provides a valuable introduction to philosophical methods of thinking. In focusing on problems about human nature, it is always clear, well-informed and challenging' – *Roger Trigg, University of Warwick*

Human Nature After Darwin is an original investigation of the implications of Darwinism for our understanding of ourselves and our situation. It casts new light on current Darwinian controversies, also providing an introduction to philosophical reasoning and a range of philosophical problems.

ISBN10: 0-415-21243-X (hbk)
ISBN10: 0-415-21244-8 (pbk)
ISBN13: 978-0-415-21243-4 (hbk)
ISBN13: 978-0-415-21244-1 (pbk)

Available at all good bookshops
For ordering and further information please visit:
www.routledge.com

RELATED TITLES FROM ROUTLEDGE

PHILOSOPHY OF SCIENCE: A CONTEMPORARY
INTRODUCTION
Alex Rosenberg

'A first-rate, challenging text that emphasizes the philosophy in the philosophy of science. Rosenberg offers a superb introduction to the epistemological and metaphysical issues at stake in modern science.' – Professor Martin Curd, Purdue University, Indiana

'Philosophy students will like the way the issues in philosophy of science are connected to the basic concerns of epistemology and philosophy of language.' – Professor Peter Kosso, Northern Arizona University

'An engaging and clearly written introduction to the philosophy of science . . . I was especially pleased to see the discussions of probability, the semantic view of theories, and science studies.' – Peter Lipton, Cambridge University

This user-friendly text covers key issues in the philosophy of science in an accessible and philosophically serious way. It will prove valuable to students studying philosophy of science as well as science students.

Prize-winning author Alex Rosenberg explores the philosophical problems that science raises by its very nature and method. He skilfully demonstrates that scientific explanation, laws, causation, theory, models, evidence, reductionism, probability, teleology, realism and instrumentalism actually pose the same questions that Plato, Aristotle, Descartes, Hume, Kant and their successors have grappled with for centuries. It features:

* chapter overviews and summaries
* a wide variety of clear supportive examples drawn from science
* study questions
* a glossary and annotated further reading

ISBN 10: 0-415-34316-x (hbk)
ISBN 10: 0-415-34317-8 (pbk)
ISBN 13: 978-0-415-34316-9 (hbk)
ISBN 13: 978-0-415-34317-6 (pbk)

Available at all good bookshops
For ordering and further information please visit:
www.routledge.com

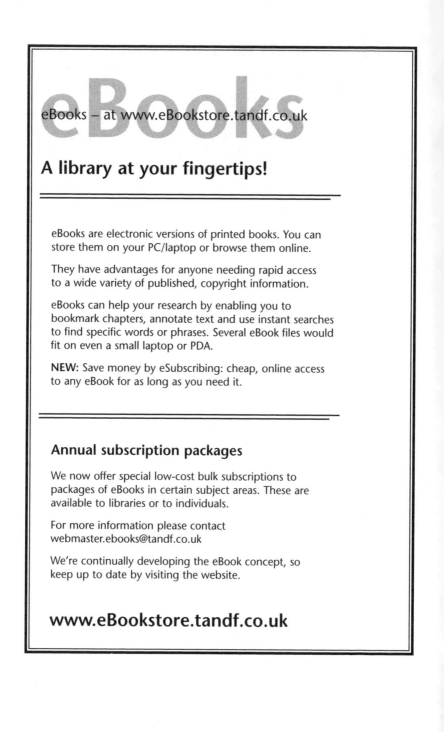

eBooks – at www.eBookstore.tandf.co.uk

A library at your fingertips!

eBooks are electronic versions of printed books. You can store them on your PC/laptop or browse them online.

They have advantages for anyone needing rapid access to a wide variety of published, copyright information.

eBooks can help your research by enabling you to bookmark chapters, annotate text and use instant searches to find specific words or phrases. Several eBook files would fit on even a small laptop or PDA.

NEW: Save money by eSubscribing: cheap, online access to any eBook for as long as you need it.

Annual subscription packages

We now offer special low-cost bulk subscriptions to packages of eBooks in certain subject areas. These are available to libraries or to individuals.

For more information please contact webmaster.ebooks@tandf.co.uk

We're continually developing the eBook concept, so keep up to date by visiting the website.

www.eBookstore.tandf.co.uk